EMBEDDED SYST...

SECOND EDITI...

Santanu Chattopadhyay

Professor
Department of Electronics and Electrical Communications Engineering
Indian Institute of Technology Kharagpur

PHI Learning Private Limited

Delhi-110092
2016

₹ 250.00

EMBEDDED SYSTEMS DESIGN, Second Edition
Santanu Chattopadhyay

ISBN-978-81-203-4730-4

The export rights of this book are vested solely with the publisher.

Fourth Printing (Second Edition) ... ... **June, 2016**

Published by Asoke K. Ghosh, PHI Learning Private Limited, Rimjhim House, 111, Patparganj Industrial Estate, Delhi-110092 and Printed by Raj Press, New Delhi-110012.

To
SANTANA, MY WIFE
My inspiration
and
SAYANTAN, OUR SON
Our hope

Contents

Preface

The second edition of the book includes more detailed information on various topics related to embedded systems. Almost all the chapters have been enriched and modified suitably by including new concepts and updated information. The book, in its second edition, comprises total 11 chapters. Chapter 2 has been elaborated by including a new *Section 2.3* which discusses the structure of ARM processor. The architectural block diagram along with the input/output pins has been presented. Discussion on THUMB instructions in *Section 2.6* has also been updated. *Digital Signal Processor* (DSP) and *Field Programmable Gate Array* (FPGA) are the other two important platforms in embedded system design. In the first edition of the book, these two topics were clubbed into a single chapter—*Chapter 3*. To provide appropriate coverage, the two topics have been divided into separate chapters—*Chapter 3* and *4* in the new edition. *Chapter 5* on "interfacing standards" has been augmented by adding various new discussions. All the interfacing standards have been updated to include recent developments in these domains. The remaining chapters have also been revised by adding explanatory notes at different places.

With this, I believe that like the previous edition of the book, the current edition will also receive wide acceptability in the academic and professional circle in the domain of embedded systems. All constructive suggestions to improve the content are welcome.

Santanu Chattopadhyay

Preface to the First Edition

Embedded Systems are foreseen to be present in almost every electrical/electronic system, in the form of computing engine embodied within them, often unnoticed by the users of the systems. Because of this, the future electronic engineers need to be equipped with the design methodology of such systems. Expertise in just only one or few domains, such as hardware, software, networking, etc. may not be sufficient to enable the designer to take wise decisions regarding the implementation platforms and design techniques to be utilized for cost-effective solutions to the design problems. An overall knowledge of all the fields with pros and cons of design alternatives is essential for designing such systems. As a subject, embedded system is an amalgamation of the fields, such as computer architecture, operating systems, modelling real-world environment, interfacing standards, networking, algorithms, and so on. This book is an effort to encompass the essentials of all these fields particularly in relation to the design of real-time embedded systems. The volume of the book has been kept to reasonable size so that the contents can be covered within a single semester. The book has been organized into ten chapters.

Chapter 1 introduces the notion of embedded system and enumerates its features to distinguish it from desktop and other computing platforms. It carefully examines the characteristics of such a system and the common design metrics. The design flow of such a system is explained, bringing out the list of tools and libraries to be used at various stages.

Chapter 2 presents one of the most widely used platforms for embedded system realization—the microcontrollers. After introducing the basic features of microcontrollers, it discusses in detail one of the most advanced embedded processor—the ARM processor. First, its history is presented. It is followed by the various attractive architectural features of ARM accounting for its popularity. Its instruction set is discussed in detail. Both ARM and THUMB instruction sets are presented. Some sample assembly language programs are also given as examples.

Chapter 3 presents an overview of other hardware platforms for embedded system realization. These include FPGA, DSP, ASIC, etc. A very detailed discussion of these is outside the scope of this book. Hence, the important features are presented for FPGA and DSP that will enable the designer to take decisions about the design platform.

Chapter 4 enumerates the interfacing standards commonly used in embedded system design. It begins with the discussion on simple strategies, such as *Serial Peripheral Interfaces* (SPI), *Inter-Integrated Circuits* (IIC), *RS-232C*, etc. The advanced versions, such as *RS-422* and *RS-485* are also presented. It is followed by a detailed discussion on *USB*. The physical, electrical and communication standards of USB are presented. Next, the wireless communication techniques, such as *IrDA* and *Bluetooth* have been dealt with. Many of the embedded applications, particularly in the automotives, utilize the *Controller Area Networking* (CAN) for information exchange between subsystems. It has also been discussed.

Chapter 5 is on specification techniques for embedded systems. This enables the readers to get familiar with the modelling techniques for the real-world systems. It begins with the discussion on StateChart, a modified version of finite state machines. Several examples have been included to illustrate the specification methodology. It is followed by another strategy, called SDL, particularly suitable for describing distributed systems. A very powerful mathematical technique for describing the behaviour of asynchronous systems is the PetriNets. The basic PetriNet structure has also been extended in several ways to handle real-time systems. All these have been discussed with a good number of illustrations. A graphical object-oriented method for embedded system specification is the UML. It consists of a set of diagrams that can be utilized to describe a system hierarchically. Suitable illustrations have been included to illustrate the specification process.

Chapter 6 deals with real-time operating systems. This forms the kernel of any moderate-to-large sized real-time embedded system. First, the tasks have been classified into several categories, such as soft, firm, and hard real-time tasks. Different scheduling algorithms have been presented, beginning with the very simple table-driven ones to the complex ones, such as rate-monotonic scheduling, earliest-deadline first scheduling, etc. The schedulability conditions for a set of tasks have been described. The pros and cons of the major scheduling strategies have been discussed. The problems related to priority-based scheduling algorithms have been presented. These include priority inversion and associated deadlock problems. The solutions to these have also been dealt with. The other general features of real-time operating systems have been enumerated. A number of such real-time operating systems have been studied and their features have been compared.

Chapter 7 presents the details of hardware–software co-simulation. It addresses the problem of verifying the correctness of the system at every stage of development. The concept of co-simulation is presented—its categories are discussed. The techniques for homogeneous and heterogeneous co-simulation have been presented. It also discusses the important issue of automated interface generation that enables the completion of hardware–software co-synthesis.

Chapter 8 deals with the issues related to hardware–software partitioning of the tasks belonging to one or more applications for a system realization. The partitioning problem takes as input a specification, most conveniently represented as a task graph with nodes representing individual tasks and the edges representing the amount of interactions and dependencies between the tasks. There exist a good number of algorithms to solve this partitioning problem targeted to architecture. These include techniques based on Integer Linear Programming (ILP), heuristic approaches (such as Kernighan-Lin algorithm), meta-search techniques (such as, Genetic Algorithm, Particle Swarm Optimization etc.). Each of these categories has been discussed. The problem has been extended further to combine with scheduling and implementation bin selection (to select among the hardware alternatives for nodes mapped onto to hardware, and similarly for software). Another important aspect of today's system design is honouring the power constraints. Some power constrained mapping solutions have also been presented.

Chapter 9 discusses the functional partitioning and optimization of task graph for an application. The problem of functional partitioning is to reconsider a given specification and come up with better grouping or refining the procedures within it so that the improved task graph can be fed as input to the partitioning process to result in better solutions. It includes steps like *granularity selection, pre-clustering* and *N-way assignment*. On the optimization side, the chapter enumerates different loop-optimization techniques, floating-to-fixed-point conversion algorithms, etc. The overall idea is to refine the manual specification into a form that can lead to a better final implementation.

Chapter 10 presents discussions on low-power techniques commonly followed in embedded system design. After presenting the basic power dissipation techniques in an electronic system, it discusses the power reduction approaches. The power can be saved at various levels—algorithm, architecture, logic, device, etc. For an embedded system designer, it is most appropriate to address the power issues at algorithm and/or architectural levels. Hence, these have been discussed in detail. The issues related to dynamic power management have been enumerated. The shutdown prediction mechanisms have been presented for periodic real-time tasks. The ACPI standards for power management have also been discussed.

This book is an attempt to bridge various domains of knowledge needed by an embedded system designer. It is estimated to be covered in a single semester undergraduate/postgraduate course on Embedded Systems. If the book can be fruitfully utilized by the students, for whom it has been written, I will consider my efforts to be successful. All constructive suggestions for improving the content will be welcomed.

Santanu Chattopadhyay

Acknowledgements

I must acknowledge the contribution of my teachers who taught me the subjects such as Digital Logic, Computer Architecture, Operating Systems, Programming Languages and Semantics, Algorithms, Networking, Compilers, VLSI Design, and so forth. The clear discussions in those classes helped me to consolidate my knowledge in these domains and combine them properly in framing the contents of the book. The various design problems introduced in the book have their roots in those class lectures. I feel deeply blessed to have such a nice group of teachers. I am also thankful to all my students, whom I have taught this subject. They all helped me to identify the gaps and mistakes in the manuscript.

My source of inspiration for writing this book is my wife *Santana*, whose relentless wish and pressure forced me to bring the book in its present shape. Over this long period, she has sacrificed a lot in the family front to allow me to have time to continue writing, taking all other responsibilities onto herself. I am also thankful to my son, *Sayantan*, for his cute comments that kept the charm of writing rolling.

I acknowledge the authors of the books and research papers, which have been referenced for writing this book. A detailed list has been provided in the Bibliography.

Thanks are also due to the publishers, PHI Learning, its editorial and production teams for providing me with the necessary support to see my thoughts in the form of a book.

Santanu Chattopadhyay

CHAPTER 1

Introduction

Information processing is the heart of any modern electrical/electronic equipment. While in eighties and early nineties, the task of information processing was accomplished via large mainframe, mini, and personal computers, the trend has changed significantly to put the computation inside the new electronic gadgets being developed in every front of life. These computational units lying totally inside the bigger device often go unnoticed by the users of the system. This continual effort to embed computational elements into bigger systems has given rise to a new class of systems, named aptly as *embedded systems*. The design goals of such systems vary significantly from the general computational systems in the sense that they often have a set of very strict performance requirements, at the same time, meeting many other design constraints. Thus, an embedded system can be defined to be a computing system embedded within a larger electronic device, performing a single (or a small set of) function(s) repeatedly, and often going unnoticed by the device's user. This definition, being a bit non-technical, may not be very precise. However, it should be noted that giving a precise definition of embedded system is quite difficult. In some sense, it can be thought to be the computing systems other than desktops and other computers with higher configurations.

Embedded systems can be found in almost every walk of life. The gadgets around us have some embedded processors in almost all of them. It is predicted that each and every electrical device will have some computational component put into it (if not already there), in the near future. Application domains of embedded systems are varied and wide. To mention a few,

- consumer electronics (cell-phone, pager, video camera, calculator, etc.)
- automobiles (anti-lock brake, fuel injection control, etc.)
- home appliances (refrigerator, washing machine, microwave oven, etc.)
- office automation equipments (fax, scanner, printer, EPBX, etc.)

In fact, the list is very long. With the newer and newer electronic gadgets reaching the market everyday, the range of applications are also increasing at a rapid pace. These systems have got varied operational principles, design philosophies, and maintenance techniques. Thus bringing these heterogeneous systems under the common umbrella of embedded systems is a challenging task. In the following section, we identify a set of features commonly found in such embedded systems.

1.1 Features of Embedded Systems

There are several features common to many embedded systems, whereas, it is also not mandatory that all features will be supported by all the embedded systems.

The following are the important features exhibited by embedded systems.

1. *Single functioned system:* Most embedded systems perform a single job repetitively. For example, a washing machine has an embedded controller that can take user inputs in terms of knob settings and perform the job of washing. A cell-phone can receive and transmit signals to enable communication between two people. A general purpose system (like a desktop) on the other hand is capable of doing a lot many operations. An embedded system will do the single function efficiently as compared to a general purpose computational system. However, it should be kept in mind that all embedded systems are not single functioned. For example, the cell-phone, as other features, may be able to send/receive SMSs, take photograph with add-on camera, tune to a radio station, play music, connect to the Internet and so on. But it cannot be utilized to perform complex scientific computation—unlike a desktop, it cannot be programmed for this purpose.

2. *Interaction with the physical environment:* Most embedded systems interact with the physical environment around them. Data are collected from the environment using sensors while actuators are used to control some of the parameters of the environment. A room temperature monitoring system may have number of temperature sensors fitted at various locations in the room. Depending upon the temperature readings, it can actuate a number heating/cooling instruments distributed in the room.

3. *User interface:* Unlike the common user interfaces like keyboard, mouse, screen, etc. in general computing systems, embedded systems often contain dedicated user interfaces consisting of push buttons, Light Emitting Diodes (LEDs), steering wheels, etc. This gives an impression of the absence of computers and thus, information processing to the general users.

4. *Dependable system:* Embedded systems are often used in safety-critical applications, like nuclear power plants, medical instrumentation, etc. This demands a high degree of dependability on such systems. It is more so, because embedded systems often work in autonomous mode, interacting with the environment and impacting upon it directly. Apart from system reliability, a dependable system must ensure easy maintainability, good availability, high degree of safety to the environment and security of information it processes. For example, a smart card reader should not release the card information to any undesired agency.

5. *Tightly constrained system:* Embedded system design is often constrained from several angles. For example, it should be a low-cost solution to the problem so that the overall system is cheap. Size of the embedded system, its performance and power budget also put severe constraint on the choice of the target implementation. For the battery to last long and reduced battery-pack size, the system must be a low-power one. These constraints, though present in other computing systems, may not be that stringent there.

6. *Real-time system:* Most embedded systems are real-time in nature. They must respond to a request from the environment within a finite and fixed amount of time. Failure to do so may lead to a catastrophic situation. For example, failure to activate fire extinguishers immediately after getting a fire alarm through sensors, may destroy the entire plant. Such systems are called *hard* real-time systems. On the other hand, if the effect is not that serious, the system is a *soft* real-time system. For example, failure to

process the image frame just arrived may create some noise in the display of the image for some time.

7. *Hybrid systems:* Many of the real-time systems are hybrid in nature, as they include both analog and digital components.

8. *Reactive systems:* Reactive systems have continual interaction with the environment. The behaviour of the system is very much dependent on the events occurring in the environment. This type of systems normally have a set of states. Depending upon the occurrence of events, state transitions in the system take place. On the other hand, a proactive system may not be interactive in nature. Once initiated, a proactive system may work on its own to produce output.

1.2 Design Metrics

The design metrics are the optimization goals that an embedded system designer wants to achieve. The commonly used metrics are the following:

1. *System cost:* It consists of two types of costs, namely the *non-recurring engineering* (NRE) cost and the recurring cost. The NRE cost is one time—the expenditure incurred in the design stage of the system. Once the system has been designed, extra units can be produced at a much lesser cost. This type of situation occurs commonly in designing VLSI chips. The NRE cost is very high as it includes the process of generating masks. However, once the mask preparation has been done, it can be replicated over a large silicon die to produce a large number of similar chips, reducing the per unit cost.

2. *Size:* Size of the system is very important. The size may be measured in silicon area for hardware, whereas it refers to the code size for the software portion of the embedded system. The code size affects the memory space requirement, increasing the overall chip/board size.

3. *Performance:* It refers to the speed of the designed system. Normally, the specification of the system will have some performance requirements to be met by the design. This is one of the vital factors influencing the decision regarding the final implementation. For example, the same functionality implemented in software will have lesser speed than a hardware realization. In the hardware realization also, an *application specific integrated circuit* (ASIC) will have better performance as compared to *Field Programmable Gate Arrays* (FPGAs) or other general purpose processors.

4. *Power requirement:* This is the other most important design metric, particularly because the embedded systems are expected to have light weight, long battery life. This necessitates plastic packaging, absence of cooling fans, etc. Thus, power requirement and the associated heat dissipation of the system should be very low.

5. *Design flexibility:* It refers to the effort needed to modify a system if the specification changes to some extent later. While a software implementation is most flexible, ASIC is the least flexible one, with FPGAs lying at an intermediary stage. The main problem in the design change is the repetition of the NRE cost which is the minimum for software.

6. *Design turnaround time:* This is the time needed to complete the design starting from specification upto taking it to the market. Due to the very high rate of obsolesence of

electronic goods, it is imperative that this time be small. The requirement often forces the designers to use off-the-shelf components, rather than doing a costly redesign of system components. Design reuse is the key term here.

7. *System maintainability:* This refers to the ease of maintaining and monitoring the health of the system after it has been put into the field. A good design is well documented such that even designers excepting those who designed the system originally, can modify the system, if necessary.

8. *Testing and verification of functionality:* It refers to the ability to check the system functionality and get confidence regarding the correct operation of it. It may be noted that in the system life-cycle, verification is generally carried out after the design has been completed. The goal of design verification is to see whether all the system features have been designed properly or not. There is no physical system available at this point of time. On the other hand, testing is needed to check for correct functioning of each unit produced. Thus, verification comes as an NRE cost while testing comes as a part of unit cost.

1.3   Embedded System Design Flow

Having understood what an embedded system is, its essential features and design parameters, now we will have a look into the embedded system design methodology. It consists of several stages as discussed next.

1. *System specification:* Design of any system starts at its specification. Specification uses a language, which may be simple English, some programming language like C (for example), or it may be some formal technique using *PetriNets, StateChart, UML chart* and so on. Different specification techniques have been detailed in Chapter 7. As will be noted later, ideally a specification should be executable, so that we can check whether the desired input–output behaviour has been modelled correctly or not. Also, it is desirable that some automated tool be able to convert the system specification into a design. These necessitate having a formal specification of the system. The type of tools doing this transformation of converting an abstract system specification into a set of sequential programs are commonly known as *system synthesis tools*. The processes do interact between themselves to realize the overall functionality of the system. The individual processes can be realized by general purpose processor or through dedicated processor. It should be noted that any task can be implemented by either type of the processors, however, the speed will vary. A general purpose processor will have a software implementation of the task, while a dedicated processor can be implemented on FPGA (Field Programmable Gate Array) or ASIC (Application Specific Integrated Circuit) to have better performance. However, the decision regarding hardware or software implementation is also often determined by the availability of pre-designed modules. Such modules form the *system level library* consisting of complete system solutions to previous problems. System specifications are normally verified by using some formal tools known as *model simulators/checkers*. These tools prepare a model of the entire system using some mathematical logic. A set of desired behaviour of the system is also specified as logic formula. The tool then checks whether those formula hold on the model or not. In case

a formula fails, the tool often generates an example of such failure. This may help the designer to rectify the error.

2. *Behavioural specification:* System specification refines to behavioural specification by the system synthesis tools. For each of the processes, a behavioural specification is obtained. As noted earlier, some of these processes are marked for software implementation on general purpose processor, while some others are on dedicated hardware. Behavioural specification is verified by *hardware–software cosimulation.* Individual simulation of only hardware or only software cannot bring out the total picture of the system. Thus, a joint simulation strategy is needed.

3. *Register transfer (RT) specification:* This is achieved through the refinement of behavioural specification. For the processes mapped onto general purpose processor, the software code is translated to the assembly/machine language instructions. It may be noted that a processor defines operation at register-transfer level only. On the other hand, for dedicated hardware realization, synthesis tools (commonly known as *high level sysnthesis tools*) convert the behavioural specification into a netlist of library components. This library includes description about RT components that may be used in the design at RT level. For example, registers, counters, ALUs, etc. The RT-specification can be verified by using RTL simulators normally available to simulate descriptions in hardware description languages, such as VHDL, Verilog, etc.

4. *Logic specification:* The specification of the dedicated processors is converted to logic specification. The logic specification consists of Boolean equations. The equations can now be converted to final implementation in some target technology. It may be noted that for the processes mapped onto general purpose processor, for which software code has been generated, no refinement is needed at this stage. Gate level simulators can be used to simulate the logic specification in terms of gates present in the circuit.

It should be noted that at each stage, it is highly necessary to verify the correctness of the refinement. This is done by simulating the specification at that stage and matching the simulated response with the desired one. At system level, a model simulator simulates the specification. It uses an abstract computational model, independent of the target processor technology, to verify the correctness and completeness of the specification. While the correctness ensures that all the desired functionalities have been specified correctly, the completeness issue ascertains that nothing extra is done by the system which may lead to undesirable system states. At the behavioural level, HDL (Hardware Description Language) simulators can be used to simulate the behaviour of the system partitions mapped onto dedicated processor. The software portion can be simulated by using a general purpose processor simulator. A cosimulator connects these two types of simulators (hardware and software) to perform the overall simulation of the system at its behavioural level. At RT-level, a structural-level HDL simulator can be used to simulate the RT-specification of the hardware components, whereas, the code corresponding to the general-purpose processor can be compiled and run. Again the co-simulator can utilize these two types of information to have a RT-level simulation of the system. At logic level, a gate-level simulator can be used to produce the output waveforms from the given input waveforms. The general-purpose processor simulators execute the program code mapped onto it. The co-simulator now checks the output of both of them to produce the final output of the system simulation. Thus the overall design methodology can be expressed in the form of a diagram shown in Fig. 1.1.

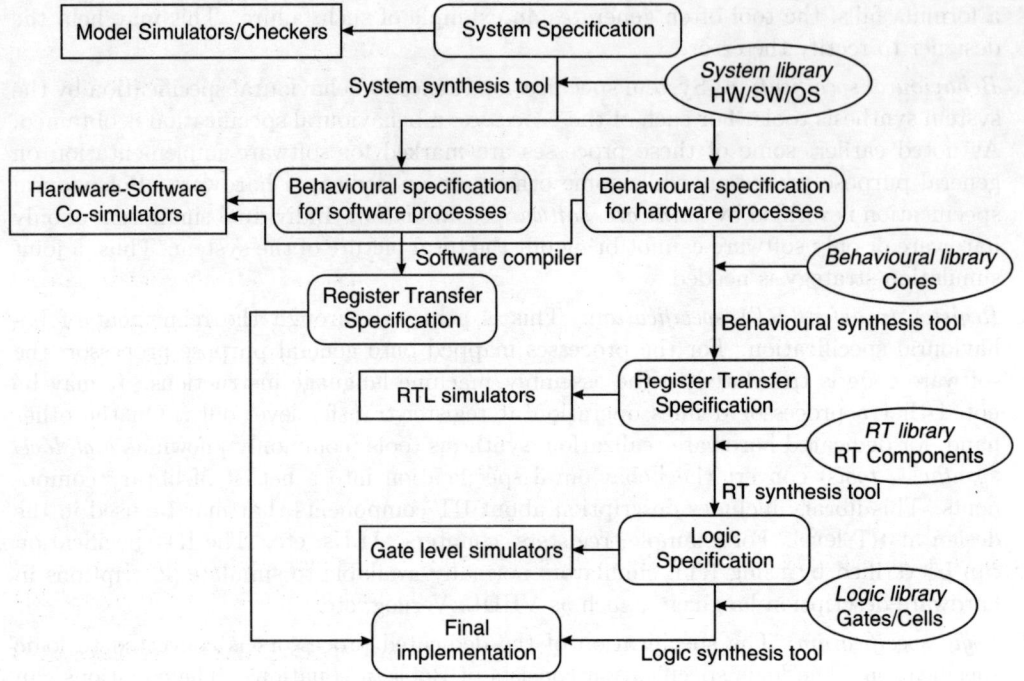

Fig. 1.1 Embedded system design methodology.

1.4   Conclusion

In this chapter, we had an overview of the embedded systems, their features, design metrics, overall design flow and the synthesis techniques. To be a successful system designer, one must have thorough knowledge about the following:

- Specification techniques to be able to specify the system behaviour in a formal fashion.

- Hardware platforms available to realize the functionalities mapped onto hardware. This requires us to know about the general purpose processors, Field Programmable Gate Arrays (FPGAs), microcontrollers, digital signal processors and so on.

- Interfaces that are commonly used, like RS232C, USB, I^2C, SPI, UART, CAN, IrDA, Bluetooth, etc.

- For efficient software design, we need knowledge about the real-time operating systems.

- Methodologies for hardware and software partitioning and synthesis.

- Automated synthesis of the interface between the system components.

In subsequent chapters we will go through discussions on each of these topics in detail. To start with, in next two chapters we will look into the available processor architectures that can be used in embedded system design.

Exercises

1.1 What is embedded system? Give as many different definitions (with justification) that you can think of.

1.2 Identify a few embedded systems around us and justify their classification as embedded system.

1.3 Differentiate between single-functioned and multi-functioned embedded systems. Give examples for each of them. How do you distinguish between a multifunction embedded system and a general desktop?

1.4 Identify a few instruments that can be part of an embedded system interacting with the environment.

1.5 Explain the terms dependability, saftey, criticality, reliability, availability with respect to an embedded system.

1.6 What is a tightly constrained system. What are the different types of constraints that we can encounter in embedded system design?

1.7 Distinguish between proactive and reactive systems. Give examples.

1.8 Explain the main features of embedded systems.

1.9 What is meant by design metric? Mention the various design metrics that need to be considered in embedded system design.

1.10 What are the components of the metric *system cost*? How does the contribution of the components change as we go from general processor-based design to application specific design?

1.11 How is the performance of a system expected to change in different implementation strategies?

1.12 What are the compelling factors that make power an important issue?

1.13 What is design turnaround time and why should it be as small as possible?

1.14 Enumerate the various steps of embedded system design.

1.15 What are the various categories of synthesis tools needed in the embedded system design cycle? Identify some potential tools (name of the tool, manufacturer, etc.) in each of theses categories.

1.16 What do you mean by the RTL specification of software processes?

1.17 What are simulators? Mention different categories of simulators in embedded system design with examples of actual tools (name of the tool, manufacturer, etc.) for each category.

CHAPTER 2

ARM: An Advanced Microcontroller

Microcontrollers are single-chip computers. In a single chip it combines a relatively simple CPU, with supports, such as timers, serial/parallel, digital/analog, input/output lines, etc. Program memory is generally included on-chip. Also, a typically small read/write memory (commonly known as *scratch-pad*) is included in the chip. To extend program and data memory further, proper interfacing facilities are provided.

While microprocessors are used in personal computers and other high-performance applications, microcontrollers are targeted towards small applications. The operating frequency may be as low as 32 kHz, though there exist many high speed microcontrollers. The major requirement of such small systems is the reduced power consumption (as noted in Chapter 1). The next important issue is of course cost. Apart from the integration of various system components into a single chip with reduced power consumption and cost, some of the important features that embedded system designers look for in microcontrollers are the following:

1. Whether the highest available speed of the microcontroller is sufficient for the application in hand.
2. The size of the chip, for example, 40-pin DIP (dual inline package), QFP (quad flat package). This determines the size of the system and thus the device.
3. Amount of on-chip ROM/RAM space should be sufficient to hold program code. Depending upon the design constraints, external memory may or may not be utilized.
4. Cost of a single chip, as it is going to determine the cost of the overall system.
5. The development platform should be good enough so that the design time is reduced. It is also advisable to have on-chip debugging facility (through JTAG port) and debug software.
6. Availability of the microcontroller chips is also another determining factor.

Various types of microcontrollers are available in the market. Some of the important ones are—68HC11, 8051, ARM, Atmel (AVR8, AVR32), Freescale (CF, S08), Hitachi (H8, SuperH), MIPS, NEC, PIC, PowerPC, TI MSP430, Toshiba TLCS-870, Zilog (eZ8, eZ80). In the following section, we look into the ARM processor architecture that has many nice features supporting embedded computation, and is, in particular, low power.

2.1 ARM Microcontroller

ARM is a 32-bit RISC (Reduced Instruction Set Computer) processor architecture developed by the *ARM Corporation*. It was previously known as *Advanced RISC Machine*, and prior to that *Acron RISC Machine*. The ARM architecture is licensed to companies that want to manufacture ARM-based CPUs or system-on-a-chip products. This enables the licensees to develop their own processors compliant with the ARM instruction set architecture. ARM processors possess a unique combination of features that makes ARM the most popular embedded architecture today.

1. ARM cores are very simple, compared to other general-purpose processors available in the market. This implies that the ARM processors will need relatively lesser number of transistors, leaving enough space on the die to realize other functionalities on the silicon.

2. The instruction set architecture and the pipeline design of ARM are aimed at minimizing the energy consumption—a critical requirement in mobile embedded systems. In spite of being a 32-bit microcontroller, it is capable of running 16-bit instruction set, known as "THUMB". This helps to achieve greater code density and enhanced power saving.

3. While being small and low-power, ARM processors provide high performance.

4. ARM architecture is highly modular—the only mandatory component of an ARM processor is the integer pipeline. All other components including caches, memory management unit (MMU), floating point and other co-processors are optional. This gives a lot of flexibility in building application specific ARM-based processors.

5. To assist the developer, the ARM core has a built-in JTAG debug port and on-chip "embedded ICE (In-Circuit Emulator)" that allows programs to be downloaded and fully debugged in-system.

2.2   A Brief History

The first ARM processor was designed by the *Acron Computers Limited* of Cambridge, England between 1983 and 1985. While looking for a new processor for the next generation desktop, they found that the existing commercial microprocessors were not suitable for their purpose, mainly because of the following two reasons.

- These processors were slower than the existing memory parts.
- Complex instruction set, including instructions requiring hundreds of cycles to execute, leading to high interrupt latencies.

Thus, the requirement of designing a new processor was felt. However, designing a complex processor used to take many years even for large companies with expertise available for processor design. The solution was found in the *Berkeley RISC 1* project which had established that it was possible to build a very simple processor with performance comparable to the most advanced CISC processors of the time. *Acron* designed their first 26-bit *Acron RISC Machine* (ARM) processor in 1985, based upon the Berkeley project. It used less than 25,000 transistors and still performed like (or better than) Intel 80286 processor that came out at about the same time. This architecture has later been referred to as *ARM version 1* architecture. It was followed by the second processor in 1987. This *ARM version 2* had a coprocessor support. It was extended with on-chip cache in *ARM 3* processor. The third version of ARM architecture, developed in 1992 had features like 32-bit addressing, support for Memory Management Unit (MMU) and 64-bit multiply-accumulate instructions. It was implemented in *ARM 6* and

ARM 7 cores. Prior to this, in 1990, *Apple* took a decision to use ARM processor in their *Newton PDA*. A joint venture called *ARM* (*Advanced RISC Machines*) was launched. ARM entered into the embedded market with the release of these processors and the *Apple Newton* PDA in 1992. The 4th generation ARM processor came out in 1996 with the special feature of THUMB—16-bit compressed instruction set. Though THUMB is slightly less efficient compared to the regular 32-bit ARM instruction set, it takes 40% less space. The most prominent representative of the 4th generation ARM is the *ARM7TDMI* core, which is till now the most popular ARM product. It has been used in most *Apple iPod* players, including the video *iPod*. Another popular implementation of *ARMv4* core is the *Intel StrongARM* processor. The 5th generation of ARM architecture introduced in 1999 has digital signal processing capability and Java byte code extensions to the ARM instruction set. *Intel XScale* processor is the most popular implementation of the 5th generation ARM core. It is used in a number of embedded devices, network processors, smart phones and PDAs. *ARMv6* architecture anounced in 2001 features improvements in many areas covering the memory system, improved exception handling and better support for multiprocessing environments. It also includes media instructions to support *Single Instruction Multiple Data* (SIMD) software execution. THUMB-2, an improved THUMB instruction set defining a new set of 32-bit instructions that execute along-side 16-bit instructions in THUMB state, was introduced. It provides better support for two separate address spaces, such that code executing in the non-secure world cannot gain access to any address space marked as secured. The protection provided by the technology is necessary for consumer privacy and extending a range of services such as, mobile banking and multimedia entertainment, to widespread consumer adoption and use. The next generation *ARMv7* cores have been introduced in 2005. They come with three different processor profiles. The 'A' profile is for sophisticated virtual memory based OS and user applications. The 'R' profile is for real-time systems, and the 'M' profile is optimized for microcontrollers and low-cost applications. The *ARMv7A* architecture has the option of *NEON* technology designed to address the next generation high performance, media intense, low-power mobile hand-held devices. It is a 64/128-bit hybrid SIMD architecture developed by ARM to accelerate the performance of multimedia and signal processing applications. The *Vector Floating Point* (VFP) coprocessor support is also an architectural option. It supports single and double precision floating point arithmetic, and is fully IEEE 754 compliant with suitable software library. Table 2.1 summarizes the discussion.

Table 2.1 ARM architecture summary

Version	Year	Features	Implementation
v1	1985	The first commercial RISC (26-bit)	ARM1
v2	1987	Coprocessor support	ARM2, ARM3
v3	1992	32-bit, MMU, 64-bit MAC	ARM6, ARM7
v4	1996	THUMB	ARM7TDMI, ARM8, ARM9TDMI, StrongARM
v5	1999	DSP and Jazelle extensions	ARM10, XScale
v6	2001	SIMD, THUMB-2, TrustZone, Multiprocessing	ARM11, ARM11 MPCore
v7	2005	Three profiles, NEON, VFP	?

2.3 Structure of ARM7

Figure 2.1 shows a block diagram of ARM7 processor. The major components of ARM7 processor are described below:

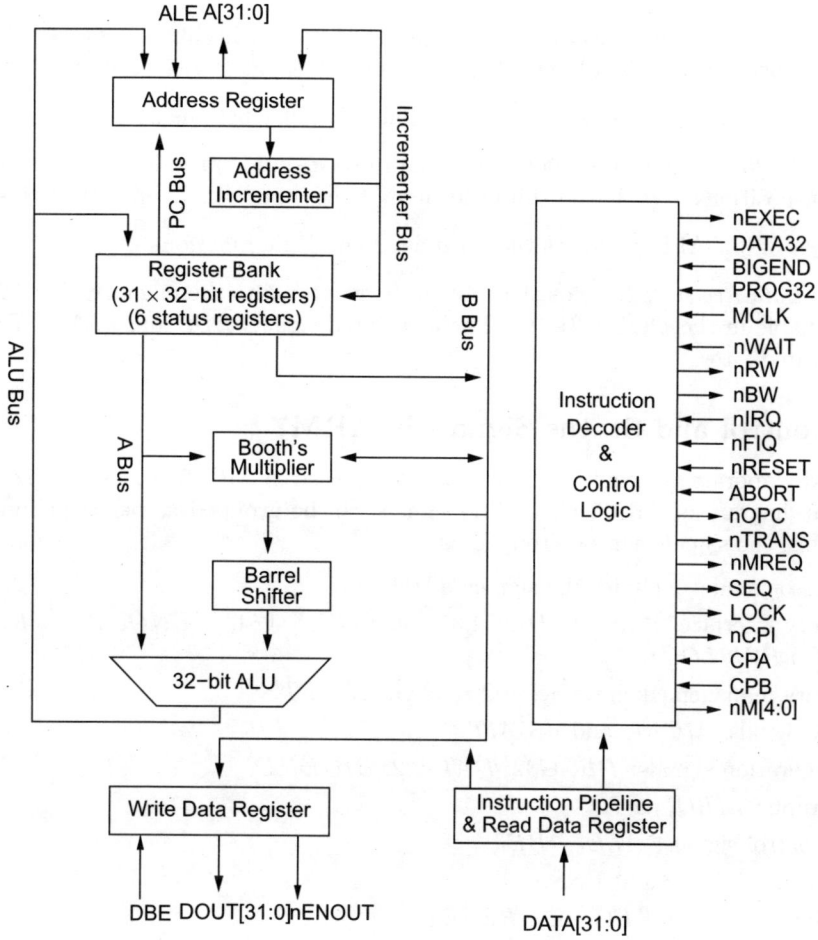

Fig. 2.1 ARM7 block diagram.

1. *Instruction Pipeline and Read Data Register:* It gets the content of memory location pointed to by the address bus lines $A[31:0]$, of *Address Register.* The external 32-bit data-in lines $DATA[31:0]$ put the content into this register.

2. *Instruction Decoder and Control Logic:* It has a number of control inputs determining the operation policy of the processor. Also, it outputs a number of control signals useful for interfacing the processor with other peripherals. The various control signals are explained later.

3. *Address Register:* It holds the address of the next instruction/data to be fetched. Address bus $A[31:0]$ originates from it. The input signal *ALE* determines the time upto which

the register's content will remain available on the $A[31:0]$ lines. Content is available as long as ALE remains low.

4. *Address Incrementer:* It increments the *Address Register*'s value by an appropriate amount to point to the next instruction/data.

5. *Register Bank:* It contains 31, 32-bit registers accessible in different modes of operation of the processor (detailed later). It also contains 6 status registers, each of size 32-bits.

6. *Booth's Multiplier:* It is used in the multiplication instructions.

7. *Barrel Shifter:* One of the operands of data processing instructions can be shifted by a few bit positions. The barrel shifter located at the input of ALU performs this function.

8. *ALU:* A 32-bit ALU performs the arithmetic and logic functions.

9. *Write Data Register:* It holds the value to be written into the memory. The 32-bit value is available in the $DOUT[31:0]$. The associated signals DBE and $nENOUT$ have been elaborated later.

2.3.1 Control and Status Signals in ARM7

We will next elaborate the control and status signals used in ARM7 processor. Fig. 2.2 shows a functional diagram of ARM7 CPU. The signals can be grouped as per their functionality. Based on this, the signals can be grouped as:

- Processor mode: includes the signals $nM[4:0]$.
- Memory interface: $A[31:0]$, $DATA[31:0]$, $DOUT[31:0]$, $nENOUT$, $nMREQ$, SEQ, nRW, nBW, $LOCK$.
- Memory management interface: $nTRANS$, $ABORT$.
- Clock signals: $MCLK$ and $nWAIT$.
- Configuration signals: $PROG32$, $DATA32$, $BIGEND$.
- Interrupts: $nIRQ$, $nFIQ$.
- Bus control signals: ALE, DBE.
- Power lines: VDD and VSS.
- Special signals: $nEXEC$ and $nRESET$.

Processor mode signals $nM[4:0]$

These status signals identify the mode of the processor. The output bits are the inverses of the internal status bits indicating processor operation mode (detailed later).

Clock signals $MCLK$ and $nWAIT$

$MCLK$ is the master clock input to the ARM processor. It has two phases. In *phase 1 MCLK* is low and in *phase 2* it is high. The clock may be stretched in either phase to interface slower devices. Alternatively, $nWAIT$ can be used. ARM7 can be made to wait for an integer number of MCLK cycles by holding $nWAIT$ low. $nWAIT$ is internally ANDed with MCLK and must only change when MCLK is low.

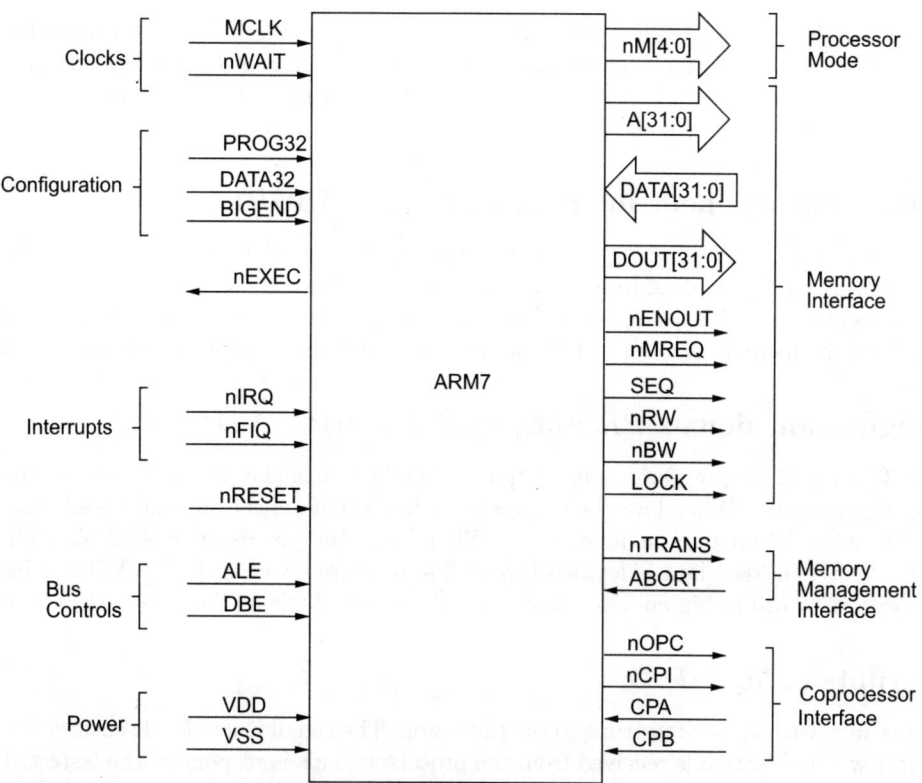

Fig. 2.2 ARM7 functional diagram.

Memory interface signals $A[31:0]$, $DATA[31:0]$, $DOUT[31:0]$, $nENOUT$, $nMREQ$, SEQ, nRW, nBW, $LOCK$

$A[31:0]$ are the address lines constituting the processor address bus. If ALE is high, address becomes valid during the phase 2 of the previous instruction cycle. This address is used in phase 1 of the referenced cycle. The stable period may be controlled by ALE as discussed later. $DATA[31:0]$ is the input data bus. During read cycles (identified by $nRW = 0$), input must be valid before the end of phase 2 of the transfer cycle. $DOUT[31:0]$ is the output data bus. During write cycles ($nRW = 1$), output data becomes valid during phase 1 and remain so for the entire phase 2 of the transfer cycle. $nENOUT$ is a status signal which is activated (made low) by the processor when $DOUT$ contains a valid data to be written into the memory. This $nENOUT$ signal can be utilized to create a bidirectional bus with $DATA$ for the memory. $nMREQ$ is a status signal indicating (when low) that the processor requires memory access during the following cycle. The signal becomes valid during phase 1, remaining valid through phase 2 of the cycle preceding that to which it refers. SEQ active-high signal indicates that the address used in the following cycle is either the same as the last memory address, or is 4 greater (i.e., the next word address). It becomes valid during phase 1 and remains valid throughout phase 2 of the cycle before the one to which it refers. The two signals $nMREQ$ and SEQ together indicate burst activity one cycle advance. nRW is another status signal.

For a read cycle, the signal is low, for a write cycle, it is high. nBW is high for a word transfer and low for a byte transfer. The signal $LOCK$ is used for locked memory access. When $LOCK$ is high, the memory controller should not allow any other device to access memory till $LOCK$ becomes low. It is used, in particular, in swap instruction.

Memory management interface $nTRANS$, $ABORT$

$nTRANS$ signal indicates when to translate the address. $nTRANS = 0$ indicates that processor is in user mode and address translation should be turned on. The timing of the signal may be modified by ALE, as discussed later. The signal $ABORT$ is an input to the processor. This allows the memory system to tell the processor that the requested access is not allowed.

Configuration signals $PROG32$, $DATA32$, $BIGEND$

$PROG32$ is for 32-bit program configuration. When high, it makes the processor to fetch from 32-bit address space. When low, the processor fetches instructions from a 26-bit address space. $DATA32$ is for 32-bit data configuration. When high, the processor uses 32-bit address for data fetch. When low, data is fetched from a 26-bit address space. $BIGEND = 1$ instructs the processor to utilize big-endian convention. When low, little-endian format is assumed.

Interrupts $nIRQ$, $nFIQ$

$nFIQ$ is an asynchronous interrupt to the processor. The signal is low-level sensitive and must be held low till an action is received from the processor. This is responded the fastest. On the other hand, $nIRQ$ is of lower priority.

Bus control signals ALE, DBE

ALE is the address latch enable, an input signal to ARM7 to extend the stable address on the address lines till end of phase 2 of MCLK. By default (ALE being 1), address changes during phase 2 of MCLK to the value needed in the next cycle. Setting ALE to 0 also extends the following signals—nBW, nRW, $LOCK$, $nOPC$, $nTRANS$. DBE is data bus enable. It facilitates data bus sharing for DMA mode of operation.

Special signals $nEXEC$ and $nRESET$

$nEXEC$ is high when the instruction in the execution unit is not being executed because, for example, it has failed its condition code check. $nRESET$ is low-level sensitive reset signal for the processor.

2.4   ARM Pipeline

One important feature of ARM architecture is its pipelined organization. There are various versions of this pipeline used in different ARM architectures. These are:

- 3-stage pipeline (ARM7TDMI and earlier).

- 5-stage pipeline (ARMS, ARM9TDMI).
- 6-stage pipeline (ARM10TDMI).
- 8-stage pipeline (ARM11).

2.4.1 3-stage Pipeline

The 3-stage pipeline is the classical *fetch–decode–execute* pipeline. The first pipeline stage reads an instruction from memory and increments the value in the instruction address register. The value is also stored in the program counter (PC). The next stage decodes the instruction and prepares control signals required to execute it on. The third stage does all the actual work: reading operands from the register file, performing ALU operations by fetching or writing to the memory the temporary data (if necessary), and finally writing back the modified register values. In the case of data processing instructions, the result from ALU is directly written into the register file, thus completing the execution stage of the instruction in one cycle, while for load/store type of instructions, the address computed by the ALU is placed on the address bus and the actual memory access is performed during the second cycle of the execution stage.

2.4.2 5-stage Pipeline

The problem with 3-stage pipeline is the pipeline stall caused by every data transfer instruction—the next instruction cannot be fetched while memory is being read/written. To circumvent this problem, in ARM9TDMI and later architectures, instruction and data memory have been separated. To make the pipeline balanced, the following measures have been taken.

- The register read step is moved to the decode stage.
- Execute stage is split into three stages. The first stage performs arithmetic computations, the second stage performs memory access, while the third stage writes the result back to the register file. It may be noted that the second stage will remain idle when executing data processing instructions.

This modification balances the pipeline, reducing the CPI (average number of *Clocks Per Instruction*). However, there is a new complication—we need to forward data between pipeline stages to resolve data dependencies between the stages without stalling the pipeline.

2.4.3 6-stage Pipeline

In ARM10 core, the instruction decode stage is split into two pipeline stages—the decode stage and the register stage. This creates a 6-stage pipeline. While the decode stage performs the decoding operation, register stage reads the register to be used. The major advancements introduced are in the width of the instruction and data buses, both of which are made 64-bit. Thus, fetch stage can fetch two instructions simultaneously. A static branch predictor module has been introduced. A separate adder has been introduced in the execution unit to take care of multiply–accummulate instructions.

2.4.4 8-stage Pipeline

Two major changes have been introduced in ARM11 cores creating an eight-stage pipeline.

- Shift operation has been separated into a separate pipeline stage.
- Both instruction and data accesses are distributed across two pipeline stages.

The execution unit is split into three different pipelines that can operate concurrently and commit instructions out-of-order also.

2.5   Instruction Set Architecture (ISA)

ARM, in most respects, is a typical RISC architecture. However, several enhancements to it have been introduced to improve the performance further. The RISC features present in ARM are as follows:

- Large uniform register file with 16 general-purpose registers.
- Load/store architecture. The instructions that process data operate only on registers and are separate from instructions that access memory.
- Simple addressing modes.
- Uniform and fixed-length instruction fields. All ARM instructions are 32-bit long and most of them have a regular three-operand encoding.

These features help in the implementation of pipelining in the ARM architecture. However, in order to keep the architecture simple and improve performance, a number of other (non-RISC) features have been introduced.

- Each instruction controls the ALU and the shifter. Thus, making the instructions more powerful.
- Auto-increment and auto-decrement addressing modes have been incorporated. This increments or decrements the value of an index register while a load or store operation is in progress.
- Multiple load/store instructions that allow to load or store up to 16 registers at once have been introduced. While violating the one cycle per instruction principle, they significantly speed up performance-critical operations, such as procedure invocation and bulk data-transfers, and lead to more compact code.
- Conditional execution of instructions has been introduced. In the machine code of an instruction, opcode is preceded by a 4-bit condition code. For the instruction to execute, the condition stated must be met. This goes a long way to eliminate small branches in the program code and eliminating stalls in the pipeline.

All these features have resulted in high performance, low code size, low power consumption, and low silicon area in ARM.

2.5.1   Registers

The ARM ISA has 16 general-purpose registers, R0–R15, in the user mode. Out of these, register R15 is the program counter which may also be manipulated as a general-purpose register. Registers R13 and R14 also have special functions; R13 is used as the stack pointer, though this has only been defined as a programming convention. Unusually, the ARM instruction set does not have PUSH and POP instructions, so stack handling is done via a set of instructions that allow loading and storing multiple registers in a single operation. R14 has special significance and is called the *link register*. When a procedure call is made, the return address is automatically placed into this register (and not in the stack, as usually done in other processors). A return from the procedure can thus be implemented by moving the content of R14 to R15. Another important register, the *current program status register* (CPSR) contains four 1-bit condition flags (namely, *negative, zero, carry,* and *overflow*) and four fields representing the execution state of the processor. The 'I' and 'F' flags enable normal and fast interrupts, respectively. The 'T' field is used to switch between ARM and THUMB instruction sets. The *mode* field selects one of the six execution modes as follows:

1. *User mode* is used to run the application code. Once in user mode, the CPSR cannot be written to. Mode can only be changed when an exception is generated.

2. *Fast interrupt processing mode* (FIQ) supports high speed interrupt handling. Generally it is used for a single critical interrupt source in a system.

3. *Normal interrupt processing mode* (IRQ) supports all other interrupt sources in a system.

4. *Supervisor mode* (SVC) is entered when the processor encounters a software interrupt instruction. These are standard ways to invoke operating system services. Upon reset, ARM enters into this mode.

5. *Undefined instruction mode* (UNDEF) is entered if the fetched opcode is not an ARM instruction or a coprocessor instruction.

6. *Abort mode* is entered in response to memory fault, for example, an instruction or data fetched from an invalid memory region.

The user registers R0 to R7 are common to all operating modes. However, FIQ mode has its own R8 to R14 registers that replace the user registers when FIQ mode is entered. Similarly, each of the other modes have their own R13 and R14 registers so that each mode has its own stack pointer and link register. The CPSR is also common to all the modes. However, in each of the exception modes, an additional register—the *saved program status register* (SPSR) is added. SPSR registers store a copy of the value of the CPSR register before an exception was raised. Figure 2.3(a) shows all the user accessible registers while Fig. 2.3(b) shows the structure of the CPSR.

Data types

The ARM instruction set supports six different data types, namely, 8-bit signed and unsigned, 16-bit signed and unsigned, 32-bit signed and unsigned. The ARM processor instruction set has been designed to support these data types in *little-* or *big-endian* format. However, most of the ARM silicon implementations use the *little-endian* format.

In the following we give a brief overview of different types of ARM instructions. ARM has got two instruction sets:

- ARM:
    - Standard 32-bit instruction set
    - It consists of the following types of instructions as shown in Fig. 2.4.
        * Data processing

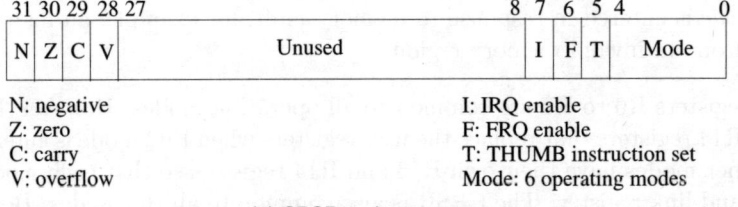

User	FIQ	Supervisor	Abort	IRQ	Undefined
R0	R0	R0	R0	R0	R0
R1	R1	R1	R1	R1	R1
R2	R2	R2	R2	R2	R2
R3	R3	R3	R3	R3	R3
R4	R4	R4	R4	R4	R4
R5	R5	R5	R5	R5	R5
R6	R6	R6	R6	R6	R6
R7	R7_fiq	R7	R7	R7	R7
R8	R8_fiq	R8	R8	R8	R8
R9	R9_fiq	R9	R9	R9	R9
R10	R10_fiq	R10	R10	R10	R10
R11	R11_fiq	R11	R11	R11	R11
R12	R12_fiq	R12	R12	R12	R12
R13	R13_fiq	R13_svc	R13_abt	R13_irq	R13_und
R14	R14_fiq	R14_svc	R14_abt	R14_irq	R14_und
R15 (PC)	R15 (PC)	R15 (PC)	R15 (PC)	R15 (PC)	R15 (PC)

User	FIQ	Supervisor	Abort	IRQ	Undefined
CPSR	CPSR	CPSR	CPSR	CPSR	CPSR
	SPSR_fiq	SPSR_svc	SPSR_abt	SPSR_irq	SPSR_und

(a) ARM registers in different modes

31	30	29	28 27		8	7	6	5	4		0
N	Z	C	V	Unused		I	F	T	Mode		

N: negative	I: IRQ enable
Z: zero	F: FRQ enable
C: carry	T: THUMB instruction set
V: overflow	Mode: 6 operating modes

(b) CPSR register content

Fig. 2.3

* Data transfer
* Block transfer
* Branching
* Multiply
* Conditional
* Software interrupts

• THUMB:

   − 16-bit compressed form
   − Code density better than most CISC
   − Dynamic decompression in pipeline

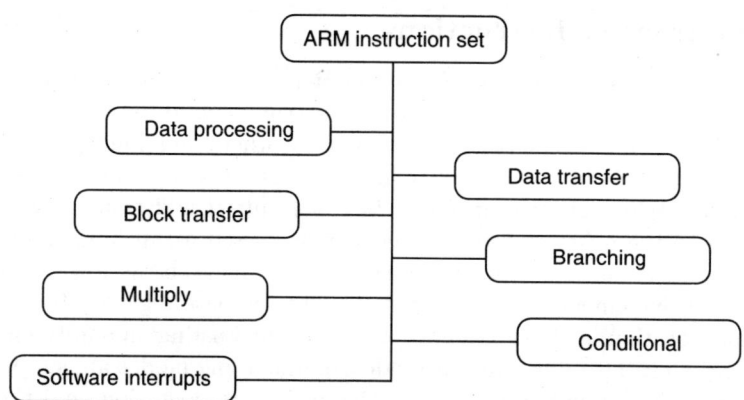

Fig. 2.4 ARM instruction set.

2.5.2 Data Processing Instructions

The ARM architecture provides a range of arithmetic operations, such as addition, subtraction, multiplication, etc., and a set of bit-wise logical operations. All these instructions take two 32-bit operands and return a 32-bit result. The multiplication instruction can return a 32- or 64-bit value. All these operands and results can be specified independently. Out of the three, the first operand and the result must be registers, while the second operand can be either a register or an immediate value. If the second operand is a register, it can be shifted or rotated before being sent to the ALU. On the other hand, for immediate operand, it must be a 32-bit value. However, all 32-bit constants cannot be specified here. This is due to the limited space available for operand specification inside the 32-bit instruction. *An immediate operand should be a 32-bit binary number where all the binary 1s fall within a group of eight adjacent bit positions on a 2-bit boundary.* More formally, a valid immediate operand n satisfies the following equation:

$$n = i \text{ ROR } (2 * r)$$

where i is a number between 0 and 255 (inclusive), r is between 0 and 15 (inclusive) and ROR is the rotate right operation. Examples of such numbers are 255 ($i = 255$, $r = 0$), 256 ($i = 1$, $r = 12$), hexadecimal number FF000000 ($i = 255$, $r = 4$), etc.

One interesting feature of the ARM architecture is that the modification of condition flags by arithmetic instructions is optional. This adds flexibility to the programming in the sense that flags do not need to be checked right after the instruction that set them, but can be done later in the instruction stream, provided that other intermediate instructions do not change the flags.

Some examples of data processing instructions are,

ADD R1, R2, R3; $R1 = R2 + R3$.
ADD R1, R2, R3, LSL #2; $R1 = R2 + (R3 \times 4)$.
ADDS R1, R2, R3, LSL #2; $R1 = R2 + (R3 \times 4)$ and set condition code flags.

2.5.3 Data Transfer Instructions

ARM supports two types of data transfer instructions: *single register transfers* and *multiple register transfers*. Single register transfer instructions can be used to transfer 1, 2, or 4-bytes of data between a register and a memory location. On the other hand, multiple register load/store operations can be carried out via multiple register transfer instructions. The addressing mode to be used for multiple register transfer is *base–plus–offset* addressing. Value in the base register is added to the offset stored in a register or passed as an immediate value to form the memory address. An *auto-indexed* addressing mode can also be used which writes the value of the base register incremented by the offset back to the base register. It helps in accessing the next memory location in the next instruction without wasting an additional instruction to increment the register. Two different auto-indexed addressing modes are supported—the *pre-indexed* mode uses the computed address for the load/store operation, and then updates the base register to the computed value. The *post-indexed* mode uses the unmodified base register for the transfer and then updates the base register. Multiple register transfer instructions also support auto-indexed addressing. Multiple register transfer instructions are particularly useful while entering or exiting a procedure to pass parameters or return values. The following points may be noted regarding the *offset* and the *indexing*.

- The offset can be
  - An unsigned 12-bit immediate value (that is, 0 to 4095 bytes)
  - A register, optionally shifted by an immediate value
- Either added or subtracted from the base register
  - Prefix the offset value or register with '+' (default) or '−'
- Applied
  - before the transfer is made: **Pre-indexed addressing** (optionally auto-incrementing the base register by postfixing the instruction with an '!')
  - after the transfer is made: **Post-indexed addressing**, causing the base register to be auto-incremented.

Some examples of data transfer instructions are,

LDR R0, [R8]	load content of memory location pointed to by $R8$ into $R0$
LDR R0, [R1, −R2]	load content of memory location pointed to by $R1 - R2$ into $R0$
LDR R0, [R1, +4]	load content of memory location pointed to by $R1 + 4$ into $R0$
LDR R0, [R1, +4]!	load content of memory location pointed to by $R1 + 4$ into $R0$, $R1$ is also incremented by 4
LDR R0, [R1], +16	Loads $R0$ from memory location pointed to by $R1$, then adds 16 to $R1$

LDR	Load word	STR	Store word
LDRH	Load half word	STRH	Store half word
LDRSH	Load signed half word	STRSH	Store signed half word
LDRB	Load byte	STRB	Store byte
LDRSB	Load signed byte	STRSB	Store signed byte

As discussed earlier, ARM supports both little-endian and big-endian formats for data access. In the little-endian format, the least significant byte of a word is stored in bits 0–7 of an addressed word, whereas, in the big-endian format, the least significant byte is stored in bits 24–31. It may be noted that this has got significance only when the data is stored as words and then accessed as bytes or halfwords. Figure 2.5 gives an example of the same.

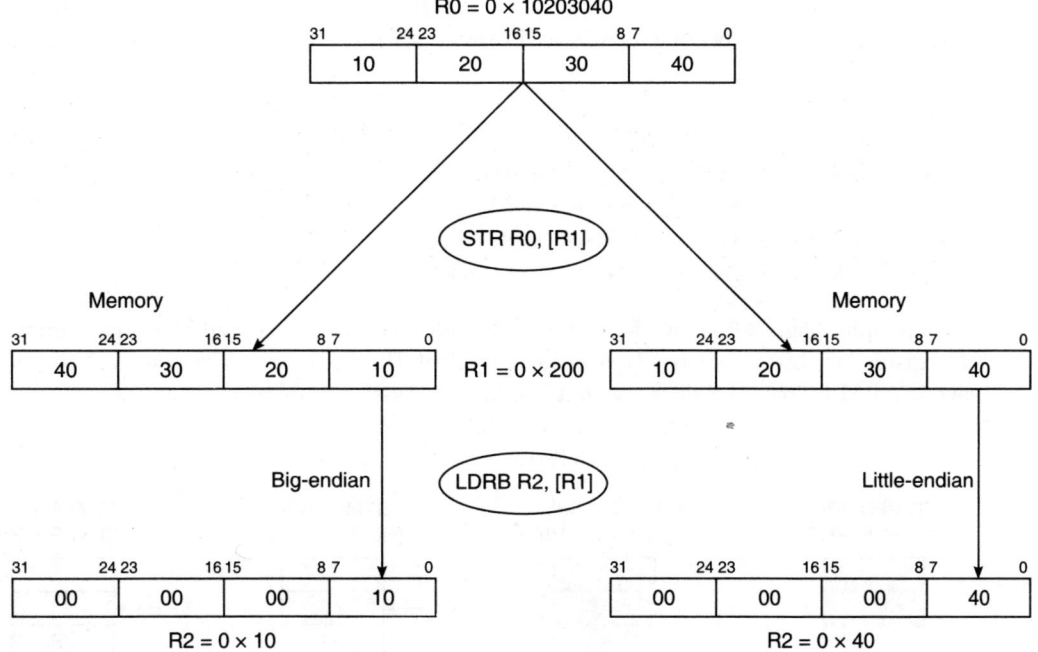

Fig. 2.5 Big-endian vs. Little-endian.

Block data transfer

The *Load* and *Store* multiple instructions (LDM/STM) allow between 1 and 16 registers to be transferred to or from memory. The transferred registers can be either:

- Any subset of the current bank of registers (default)
- Any subset of the user mode bank of registers when in a privileged mode (postfix instruction with a ˆ symbol).

Like single register transfer operations, in this case also the *base register* is used to determine where memory access should occur. Auto-increment, and auto-decrement are also supported for the base register. Lowest register number is always transferred to/from the lowest memory location accessed. The block transfer instructions have got efficient utilization in

- implementing stack for saving and restoring context
- moving large blocks of data around memory

1. **Stack operation:** Though traditionally a stack grows down in memory with the last pushed value at the lowest address, ARM can be made to implement "ascending stack" also, where the stack grows up through memory. The value of the stack pointer can be either:

   - Point to the last occupied address (Full stack), and thus needs pre-decrementing before the push
   - Point to the next occupied address (Empty stack), and thus needs post-decrementing after the push

The stack type to be used is given by the postfix to the instruction as follows:

   - STMFD/LDMFD: Full descending stack
   - STMFA/LDMFA: Full ascending stack
   - STMED/LDMED: Empty descending stack
   - STMEA/LDMEA: Empty ascending stack

For example, Fig. 2.6 shows four stack examples. It may be noted that the sequence of registers within an instruction does not affect the order in which the registers are saved. ARM follows the strategy that the lowest register number is always stored at the lowest address. Thus, instead of writing the sequence as $R0, R1, R3 - R5$ specifying it as $R1, R0, R5, R4, R3$ or $R5, R0 - R1, R3 - R4$ have the same effect.

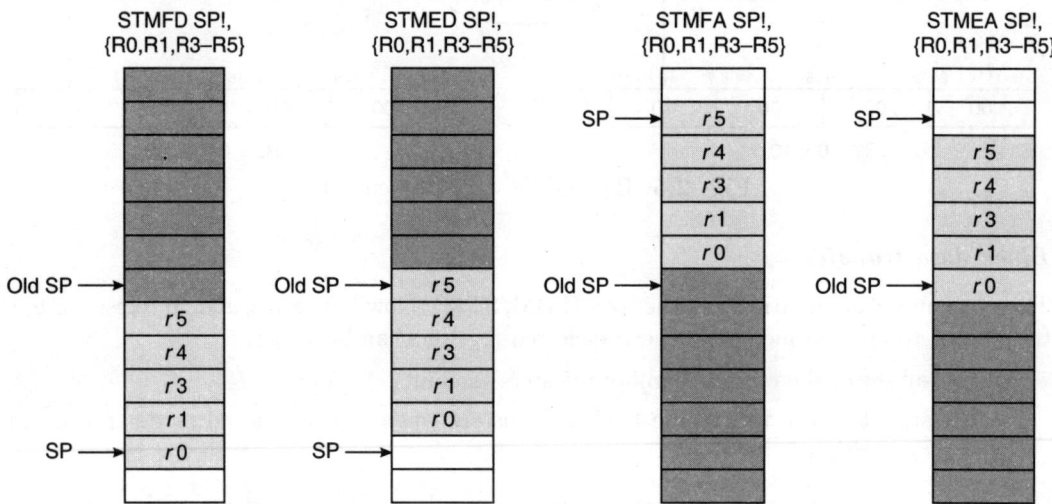

Fig. 2.6 Example stack.

2. **Moving a large data block:** We can best explain it with the help of an example. Suppose, we have to copy a block of memory which is an exact multiple of 12 words long from the location pointed to by register $R12$ to the location pointed to by $R13$. $R14$ points to the end of the block to be copied. The following versions of LDM/STM instructions can be utilized for this purpose:

   - STMIA/LDMIA: Increment after

- STMIB/LDMIB: Increment before
- STMDA/LDMDA: Decrement after
- STMDB/LDMDB: Decrement before

The exact code to perform the task may be as follows:

```
; R12 points to the start of the source data
; R14 points to the end of the source data
; R13 points to the start of the destination data
loop        LDMIA R12!, {R0-R11}      ; load 48 bytes
            STMIA R13!, {R0-R11}      ; and store them
            CMP R12, R14              ; check for the end
            BNE loop                  ; and loop until done
```

2.5.4 Multiplication Instructions

ARM provides several versions of multiplications. These are:

- Integer multiplication (32-bit result)
- Long integer multiplication (64-bit result)
- Multiply accumulate instruction—particularly useful in signal processing applications, for example, digital filtering.

Accordingly, there are several versions of the multiplication instruction.

MUL	Multiply	32-bit result
MULA	Multiply accumulate	32-bit result
UMULL	Unsigned multiply	64-bit result
UMLAL	Unsigned multiply accumulate	64-bit result
SMULL	Signed multiply	64-bit result
SMLAL	Signed multiply accumulate	64-bit result

For example,

MUL $R0$, $R1$, $R2$ $R0$ gets $R1 * R2$
MULA $R0$, $R1$, $R2$, $R3$ $R0$ gets $R1 * R2 + R3$

However, there are a few restrictions regarding the source and destination.

1. Destination and the first operand cannot be in the same register.
2. PC (R15) cannot be used for multiplication.

ARM uses the Booth's Algorithm to perform integer multiplication. In some variants of ARM processors, multiplication proceeds as follows:

- For each pair of bits, it takes 1 cycle. One more cycle is needed to start the instruction. The multiplication continues till the source register has some 1's left in it. Otherwise, the algorithm *early-terminates*. For example, to multiply 18 (00000000000000000000000000000000 10010 in binary) and -1 (1111111111111111111111111111111 in binary), if the source register content is 18, it will take 4 cycles, whereas, if the source register holds -1, number of cycles needed is 17. Generally, the language compilers take care of such issues.

In some other variants containing extended multiplication hardware, the following enhancements have been performed.

- An 8-bit Booth's algorithm is used, which makes the multiplication faster (maximum standard instruction is now 5 cycles).
- Early termination method is improved in the sense that the multiplication completes when all the remaining bit sets contain either all 0s, or all 1s.
- 64-bit results can be produced from 32-bit operands, this provides higher accuracy. A pair of registers is used to hold the result.

2.5.5 Software Interrupt

The software interrupt (SWI) instruction forces the CPU into supervisor mode. Its format is,

$$\text{SWI} \ \#n$$

The execution of the instruction causes an exception trap to the SWI hardware vector (forcing the mode change and an associated state saving), thus causing the SWI exception handler to be called. The handler can now analyse the value of n to determine which action to perform. It should be noted that the processor completely ignores n, and it is the software interrupt handler that may analyse the value of n to perform the desired job. This is very suitable for an operating system to implement a set of privileged operations which applications running in user mode can request. Such requests are commonly known as *system calls*. The value of n is 24-bits, thus providing facility for a maximum of 2^{24} calls. Unlike many other processors (such as, Intel processors), the hardware does not attempt to distinguish between these 2^{24} cases. If separate addresses are to be maintained for them, a total of $2^{24} \times 4$ (address bus is 32 bits = 4 bytes) = 2^{26} bytes of memory will be needed.

2.5.6 Conditional Execution

This is an interesting feature of ARM instruction set. While most of the existing architectures allow only branches to be executed conditionally, ARM allows *all* instructions to be executed conditionally. The most significant four bits of each instruction are reserved to hold the 16 condition codes. As it may be noted, there are four flags $N, Z, C,$ and V. The following are the interpretations of these four-bit settings.

0000	: EQ – (Equal)
0001	: NE – (Not equal)
0010	: HS/CS – (Unsigned higher or same)
0011	: LO/CC – (Unsigned lower)
0100	: MI – (Negative)
0101	: PL – (Positive or zero)
0110	: VS – (Overflow)
0111	: VC – (No overflow)
1000	: HI – (Unsigned higher)
1001	: LS – (Set unsigned lower or same)
1010	: GE – (Greater or equal)

1011	: LT – (Lower)
1100	: GT – (Greater)
1110	: AL – (Always)
1111	: NV – (Reserved)

We have already seen how to set the conditional flags in instruction execution. For example, the use of mnemonic *ADDS* in place of *ADD* sets the flags, whereas the basic *ADD* instruction does not affect any of the flags. Now, to execute an instruction conditionally, we have to use mnemonics like *EQADD*. For example:

EQADD R0, R1, R2   perform R0 = R1 + R2 only if zero flag is set.

EQADDS R1, R2, R3, LSL #2   perform R1 = R2 + (R3 × 4) only if zero flag is set, and set condition code flags.

2.5.7 Branch Instruction

In ARM processor, the following features of branch instructions can be noted:

1. All branches are relative to the program counter.
2. Jump is always within a limit of ±32 MB.
3. Conditional branches are formed by using the condition codes as discussed earlier.
4. Subroutine call instruction is also modelled as a variant of branch instruction.

There are two opcodes reserved for branching—**B** (standard branch) and **BL** (branch with link, current value of PC+4 is saved in link register R14). The **BL** version can be used to call subroutine. The return from subroutine can be affected by copying the content of R14 back to PC. The branch instruction structure is as shown in Fig. 2.7.

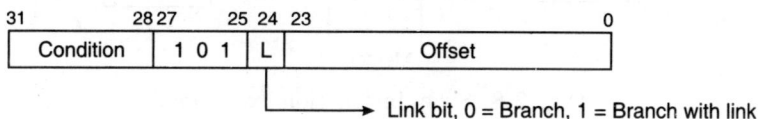

Fig. 2.7 Branch instruction format.

In the coding of a branch instruction, to get the offset, first a 26-bit difference is computed between the branch instruction and the target. It is then right shifted by two bits (as the instructions are always word-aligned, least significant two bits are always 0s), and stored in the instruction encoding. This provides a branch range of ±32 MB. While executing the branch, the processor shifts the offset left by two bits, sign extends it to 32-bits, and adds it to the PC to get the branch target. The instruction pipeline is filled by getting instructions from the target address, and the execution resumes.

Another very important class of branch instructions is the *Branch exchange*—**BX** and **BLX**. These are similar to **B** and **BL** instructions, however, it also performs the exchange of instruction set between ARM instructions and THUMB instructions. **This is the only way to swap instruction sets.**

2.5.8 Swap Instruction

A careful look into the ARM instructions will reveal that a single instruction can read/write at the most one memory location. The *swap* instruction is an exception to this. *Swap* is an *atomic* operation in which a memory read is followed by a memory write which moves byte or word between registers and memory. Its format is,

    SWP Rd, Rm, [Rn]
    SWPB Rd, Rm, [Rn]

The execution proceeds as follows (as depicted in Fig. 2.8).

1. It is a two-cycle operation.
2. Content of memory location pointed to by register Rn is copied into a temporary space.
3. Content of register Rm is copied into the memory location.
4. Content of the temporary space is copied into the register Rd.

Thus, to effect an interchange between the registers Rd and Rm, they should be made the same register. In that case, content of the memory location is swapped with the register in an atomic operation. This can be exploited to implement the *semaphore* operations of the operating system.

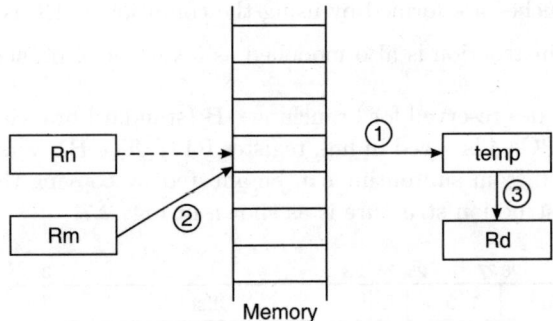

Fig. 2.8 Swap instruction execution.

2.5.9 Modifying Status Registers

The status registers (i.e., CPSR and SPSR) can only be modified indirectly. The instruction *MSR* moves content from CPSR/SPSR to the selected general purpose register, whereas the *MRS* moves content of selected GPR to CPSR/SPSR. The instruction can only be executed in the privileged modes.

2.6 THUMB Instructions

As we have noted earlier, the ARM processor supports two different instruction sets—*ARM* and *THUMB*. While the ARM instruction set has got 32-bit instructions, THUMB instructions are 16-bit in length. These instructions are stored in a compressed form. The instructions are decompressed into ARM instructions and then executed by the processor. Figure 2.9 shows the THUMB instruction processing. Figure 2.10 shows the THUMB instruction decompressor.

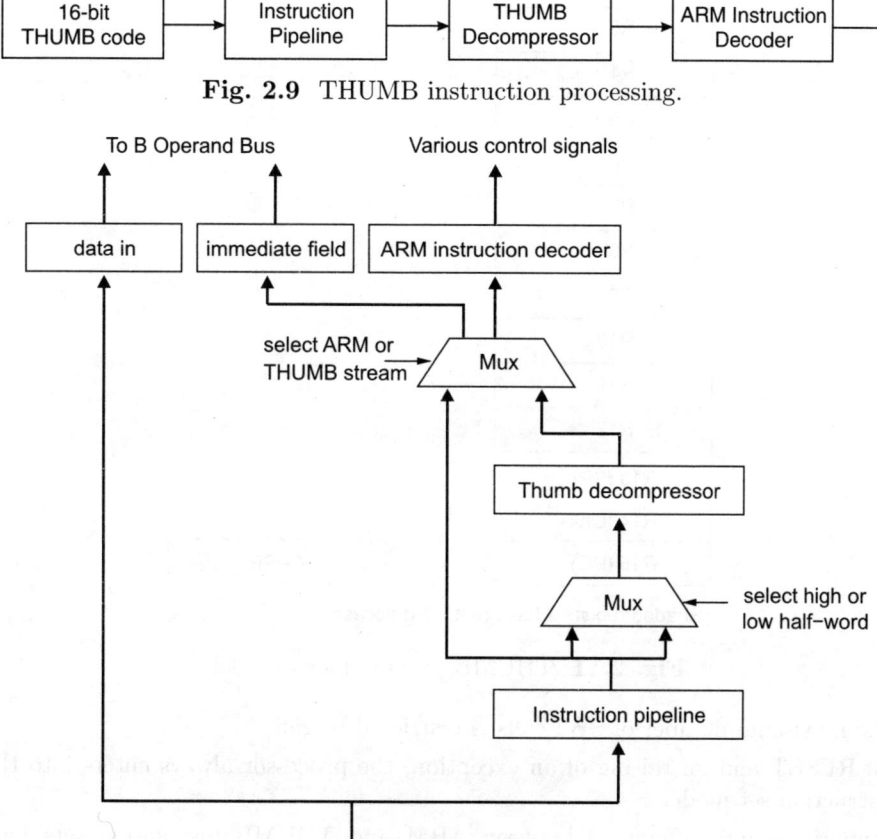

Fig. 2.9 THUMB instruction processing.

Fig. 2.10 THUMB instruction decompressor.

THUMB instruction set must always be entered by running a BX/BLX instruction. These instructions are similar to their ARM counterparts, with a few exceptions.

1. THUMB instructions are executed unconditionally, excepting the branch instructions.

2. THUMB instructions have unlimited access to registers R0–R7 and R13–R15. A reduced number of instructions can access the full register set. As shown in Fig. 2.11, only a few instructions can access registers R8–R12.

3. The instructions look more like a conventional processor's instructions. For example, instructions like PUSH and POP are present for stack manipulation, though the final implementation is via the ARM multibyte transfer instructions. It implements a descending stack with the stack pointer hardwired to R13.

4. No MSR and MRS instructions.

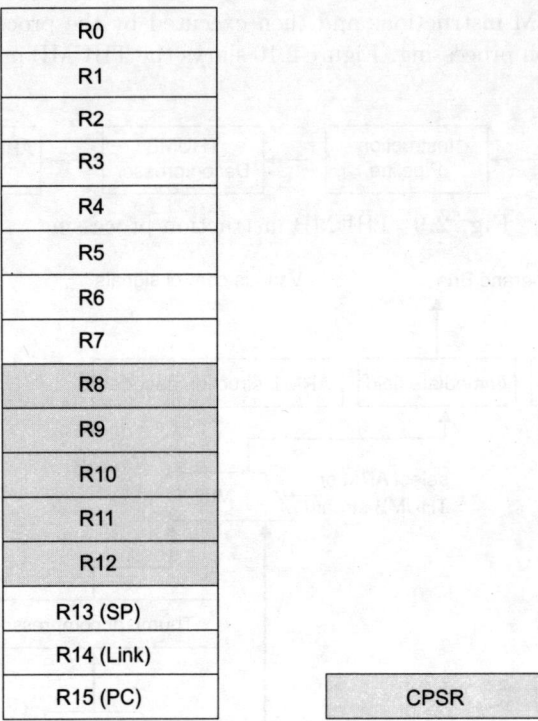

Shaded registers have restricted access

Fig. 2.11 THUMB programmer's model.

5. The maximum number of SWI calls is restricted to 256.

6. On RESET and on raising of an exception, the processor always enters into the ARM instruction set mode.

7. Similarities and differences between ARM and THUMB instruction sets have been detailed in Table 2.2.

Table 2.2 Comparison between ARM and THUMB instruction sets

Similarities	Differences
1. Load-store architecture	1. Most THUMB instructions are unconditional, while all ARM instructions are conditional
2. Support for 8-, 16- and 32-bit data types	2. Most THUMB instructions use a 2-address format, while most ARM instructions use a 3-address format
3. 32-bit unsegmented memory	3. THUMB instruction formats are less regular – a result of denser encoding
	4. THUMB has explicit shift opcodes, while ARM implements shifts as operand modifiers

The basic advantage of THUMB comes in the form of higher code density than ARM code. The THUMB code requires, on an average, 30% less space. If the memory is organized as

32-bit words, ARM code is 40% faster than THUMB. However, if the memory is organized as 16-bit words only, THUMB code is found to be faster than ARM code by 45%. The most important advantage of THUMB comes in the form of power savings. THUMB code uses up to 30% less power than ARM code. Except for the most speed-critical embedded devices, the cost of memory and power are much more critical than the execution speed of the processor. This is why the THUMB instruction set may be the choice of many simple embedded system designs. The choice between ARM and THUMB instruction sets can be made as follows:

1. For the best performance, 32-bit memory and ARM instruction set should be used.
2. For best cost and power efficiency, it is advisable to use 16-bit memory with THUMB code.
3. In a typical embedded system

    (a) ARM code should be used in 32-bit on-chip memory for small speed-critical routines.
    (b) THUMB code should be used in 16-bit off-chip memory for large non-critical control routines.

2.7 Exceptions in ARM

The exceptions can be raised in ARM in different situations. These can be split into following three different categories:

1. Exceptions can occur through execution of instructions. These include the software interrupts, undefined instructions and memory abort.
2. Exceptions can also be caused as a side-effect of an instruction, like data fetch failure.
3. Other sources can be such as RESET, FIQ, and IRQ interrupts.

For all these exceptions, the mechanism followed is similar. The processor switches to the privileged mode, current value of PC(R15) + 4 is saved into the link register (R14) of the privileged mode. Current CPSR is saved in the SPSR of the privileged mode. The IRQ interrupts are disabled. If the exception raised is an FIQ, the FIQ interrupts are also disabled. Next, the program counter is loaded with the exception vector address, and the execution of the exception starts. Table 2.3 shows the vector table for different ARM exceptions.

Table 2.3 ARM7 vector table

Exception type	Mode	Meaning
Reset	Supervisor	0x00000000
Undefined instruction	Undefined	0x00000004
Software interrupt (SWI)	Supervisor	0x00000008
Prefetch abort (instruction fetch abort)	Abort	0x0000000C
Data abort (data access memory abort)	Abort	0x00000010
IRQ (interrupt)	IRQ	0x00000018
FIQ (fast interrupt)	FIQ	0x0000001C

Two important issues to be noted here are:

1. The vector at location 0x00000014 is missing to ensure a backward compatibility.
2. The FIQ has been given the highest address, so that the interrupt service routine can start from that address directly. This eliminates the necessity of another jump instruction from the vector address to the interrupt service routine, thus saving the time needed to start the routine after the fast interrupt has occurred.

2.8   Programming Examples

In this section, we will look into a few programming examples of moderate complexity. The idea is to get an understanding about how to use the ARM instructions in an assembly language program.

2.8.1   Finding the Maximum of a Set of Numbers

Assume that the register $R0$ holds the start address of the data block (each data is 4 bytes long) and $R2$ holds the number of elements in the block. The maximum of the elements will be stored in register $R1$.

```
          EOR     R1, R1, R1      ;clear R1 to store the largest
          CMP     R2, #0
          BEQ     Over            ;if block is empty, done
Loop
          LDR     R3, [R0]        ;get the data
          CMP     R3, R1          ;do comparison
          BCC     Looptest        ;skip if R1 is bigger
          MOV     R1, R3          ;else get the larger in R1
Looptest
          ADD     R0, R0, #4      ;increment pointer R0
          SUBS    R2, R2, #1      ;decrement number of elements left
          BNE     Loop            ;if not done, loop
Over
                                  ;R1 holds the largest
```

2.8.2   Comparing Two Null-terminated Strings

In this example, we assume that the two strings are pointed to by the registers $R0$ and $R1$ respectively, both the strings are null-terminated (i.e., the last byte is 0), and that the result will be stored in register $R2$. If the two strings match exactly, the content of $R2$ will be 0, else it will be -1.

```
Loop
              LDRB    R3, [R0]           ;get next character of string 1
              LDRB    R4, [R1]           ;get next character of string 2
              CMP     R3, R4             ;compare
              BNE     Notsame            ;if not same, strings do not match
              CMP     R3, #0             ;check if end of string reached
              BEQ     Same               ;if equal, same
              ADD     R0, R0, #1         ;increment pointer to string 1
              ADD     R1, R1, #1         ;increment pointer to string 2
              BAL     Loop               ;branch always to check next character
Notsame
              MOV     R2, #-1            ;mark not matched
              BAL Over
Same
              MOV     R2, #0             ;mark matched
Over

                                         ;R2 holds the match
```

2.9   Conclusion

In this chapter, we have seen a broad overview of ARM microcontroller architecture. It has many interesting architectural features combining both CISC and RISC philosophy. The facilities provided in FIQ handling reduce response time to critical interrupts. The controlled modification of status bits by arithmetic instructions and conditional instruction execution are unique to the processor. The large number of system calls that can be supported by SWI instruction is definitely helpful for the OS designers. Availability of a large number of registers helps in speedy context switch. The highly compact and low-power THUMB instructions make the processor an ideal choice for small, low-cost embedded system development.

Exercises

2.1 How does a typical microcontroller differ from a microprocessor? What are the typical architectural blocks present in a microcontroller?

2.2 What are the factors guiding the choice of a microcontroller for an embedded application?

2.3 What is meant by 'application specific' ARM processor? How does it help the embedded system designer?

2.4 How does 'in-circuit emulation' help in system development?

2.5 Explain the internal structure of ARM processor.

2.6 How does the ALE signal of ARM differ from ALE of *x86* family of processors?

2.7 Explain the functional diagram of ARM processor.

2.8 Enumerate the evolution of various pipelining structures in ARM.

2.9 Compare and contrast between RISC and CISC instructions. Explain how in ARM both these philosophies have been merged.

2.10 How many general and special-purpose registers are there in ARM? Explain their functionalities.

2.11 State the role of R13, R14, and R15. How does R15 differ from program counter in a general CPU?

2.12 Enumerate various operating modes of ARM.

2.13 Mention different data types available in ARM. How does a big-endian type differ from little-endian type?

2.14 Suppose the number 0x12345678 is stored at memory location 1000 in big-endian format. If the processor now assumes the data to be in little-endian format, what will it get if it reads (a) a byte, (b) a half-word, and (c) a word from location 1000? In the previous example, suppose that the data bus lines are connected to the memory in the reverse order, that is bit-0 from processor is connected to data line 31 of memory, 1 to 30, and so on. What value will be read in cases (a), (b) and (c) noted above?

2.15 Enumerate the types of instructions in ARM.

2.16 Why is it the case that all possible 32-bit operands cannot be specified as immediate operands in ARM data processing instructions? Out of the following numbers, which can be represented as immediate operand: 56, 1000, 513, −1?

2.17 How to set condition flags in arithmetic operations in ARM? What advantage does this preferential setting provide us?

2.18 Mention different ways in which the address of a memory operand can be specified in ARM.

2.19 Why do you think that not providing separate stack-space in ARM may be beneficial? Show the content of the stack after executing each of these instructions on a partially filled stack – STMFD SP!, {R0-R12}, STMED SP!, {R0-R12}, STMFA SP!, {R0-R12}, STMEA SP!, {R0-R12}.

2.20 What do you mean by early-termination? If the operands for multiplication are 12 and 55, which one should be used as the source operand?

2.21 State the differences between the ways in which software interrupts are handled in Intel processors and in ARM.

2.22 Explain the conditional execution of statements in ARM. For the following piece of high-level code, show the corresponding assembly-level code for ARM. Assume all the variables are available in registers.

```
if a > b then
     x = y + z
else
     x = y - z
```

2.23 What are the advantages of program counter 'relative jump' over 'absolute jump'? Why do you think that the 24-bit "offset" in the branch instruction structure is sufficient for ±32 MB branch range in ARM? Assuming that the current program counter is at 0x10000000, compute the 24-bit offset for the addresses 0x10000020 and 0x0FFFFFFC.

2.24 How does BX/BLX differ from B/BL?

2.25 What is meant by an atomic operation? How is the SWAP instruction different from others from atomicity viewpoint? How does it help in implementing "semaphores", explain with suitable example.

2.26 What is THUMB? How does the THUMB instruction set differ from ARM instruction set? Are the THUMB instructions executed directly?

2.27 Why are the FIQ interrupts can be handled faster than IRQ interrupts in ARM?

2.28 Write the following programs in the assembly language of ARM.

(a) Sort a given set of numbers.
(b) Find second highest of a set of numbers without actually sorting it.
(c) Concatenate two null-terminated strings.
(d) Look for a specific substring within a null-terminated string.

2.29 Perform a comparative study between the microcontrollers (i) ARM, (ii) 8051, (iii) PIC, (iv) AVR.

CHAPTER 3

Digital Signal Processors

In the previous chapter, we have seen ARM processor, a representative from the category of microcontrollers. It has got many nice features that make it very attractive for low-cost embedded system design. However, where the performance becomes a very important criteria, we have to look for alternatives. Since many of the embedded systems interact with nature and environment, the input signals are predominantly analog. For example, the image and speech signals. Such signals require special type of processing to extract the core information part from it (since, the received signal may have good amount of noise associated with it), modify the signal for a particular application and then transmit it to the output devices/environment. Though a general-purpose processor can cater to this type of applications also, another class of special-purpose processors have been designed for these applications. Such a processor is broadly known as a *Digital Signal Processor* (DSP). Compared to a general-purpose microprocessor/microcontroller, signal processing applications can better be run on DSPs to have low-cost, higher performance, lower latency solutions. It also often removes the requirements of specialized cooling and larger battery sizes. This makes DSPs, particularly suitable for portable embedded applications, such as mobile phones and PDAs.

Figure 3.1 shows a typical digital signal processing system. The input analog signal is converted into its digital form through an *Analog-to-Digital Converter* (ADC). This digital signal is now used by the DSP to perform the signal processing job. The output digital signal from the processor is converted back to its analog form by using a *Digital-to-Analog* (DAC) converter, which outputs the analog signal.

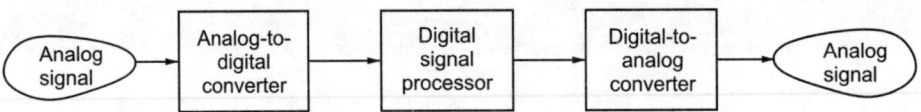

Fig. 3.1 Typical digital signal processing system.

Digital Signal Processing involves a set of mathematical operations carried out on a sequence of input data/signal, such as digital filtering, Fourier analysis etc. *Digital Signal Processors* are microprocessors, specifically designed to efficiently handle digital signal processing tasks, unlike general processors that can perform data manipulation (such as, word processing, database management etc.), over and above general arithmetic and logic operations. However, it is difficult to make a device that is optimized for both classes of applications. While a data manipulation may involve storing and retrieving data to/from memory locations, and do some limited amount of computation involving them, digital signal processing is highly computationally intensive. For example, an FIR filter calculates the sample at location n in the output signal $y[n]$ as,

$$y[n] = a_0 x[n] + a_1 x[n-1] + a_2 x[n-2] + \ldots$$

That is, the input signal has been convolved with a filter kernel consisting of $a_0, a_1, \ldots$. The kernel may have only a few coefficients, or many thousands, depending upon the application. The overall FIR computation time depends upon the number of multiplication and addition operations, rather than data transfer and loop termination checks, etc. In most of the applications of DSP, the processor must have a predictable execution time. In many cases, it is used in real-time environment that continually processes input signals. Thus, the architecture of a DSP processor is expected to differ significantly from general purpose processors.

3.1 Architecture of Digital Signal Processors

Most of the DSP processors have the ability to perform one or more *multiply-accumulate* (MAC) operations as single instructions. This is very suitable for many signal processing applications. For example, consider the following computation.

$$y_{i,j} = y_{i,j-1} + x_{i-j} * a_j$$
$$y_{i,-1} = 0$$
$$y_i = y_{i,n-1}$$

It can be implemented efficiently by the MAC unit as shown in Fig. 3.2.

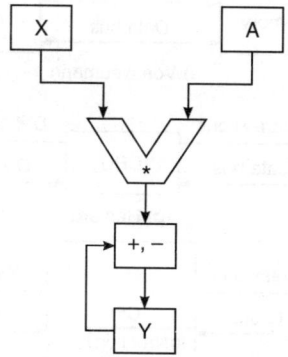

Fig. 3.2 Typical multiply-accumulate unit.

DSP processors are capable of completing several accesses to memory in a single instruction cycle. Thus, at the same time, the processor can fetch data, such as samples of the input signal and filter coefficients, as well as program instructions. To support simultaneous access to multiple memory locations, DSP processors provide multiple on-chip buses, multi-ported on-chip memories, and in some cases multiple independent memory banks.

DSP architecture design is guided by a number of special requirements for high throughput signal processing tasks. The following are some of the salient features of a DSP core:

1. Fast data access through high bandwidth memory architecture, specialized addressing modes, direct memory access (DMA) etc.

2. Fast computation via MAC unit, pipelining, parallel architecture including VLIW (Very Long Instruction Word) and SIMD (Single Instruction Multiple Data).

3. Higher accuracy obtained via extra-wide accumulator, guard bits etc.

4. Faster execution control using hardware assisted zero-overhead loops, shadow registers, etc.

3.2   High-speed Data Access

Since off-chip memory access is the most time consuming operation in program execution, rate of transfer of data to/from memory has been improved in several ways in DSPs. The strategies are noted in the following subsections.

3.2.1   Memory Architecture Design

Figure 3.3(a) shows the memory access for a traditional microprocessor. This is known as *Von Neumann* architecture. It has a single memory and a single bus transferring data into and out of central processing unit. In contrast, DSP processors typically use *Harvard* architecture (Fig. 3.3(b)) or *Super Harvard* architecture (Fig. 3.3(c)). *Harvard* architecture possesses separate program and data memory. While program memory contains instructions, data memory holds data. The two memory modules have separate address and data buses. The *Super Harvard* architecture adds the features like *on-chip instruction cache* in the CPU and *I/O controller*.

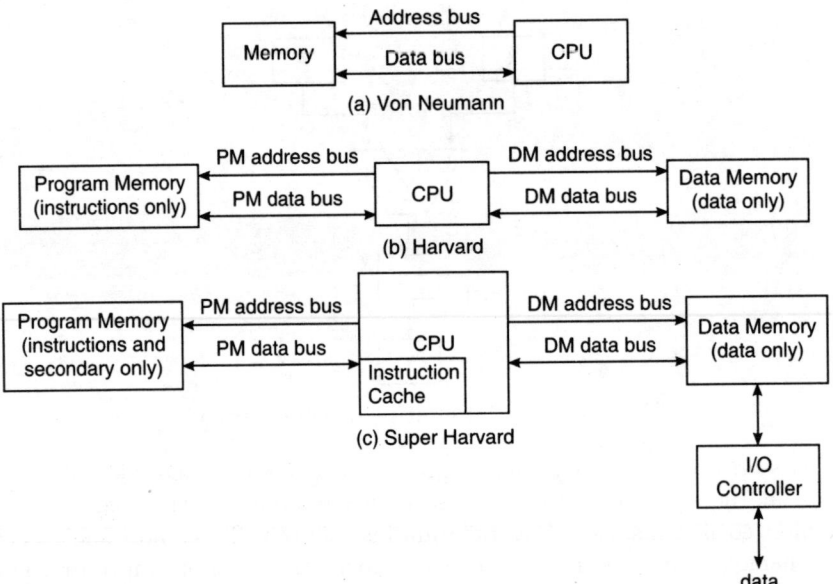

Fig. 3.3 Various memory-processor interaction architectures (a) Von Neumann (b) Harvard (c) Super Harvard.

To differentiate between the performance of the three architectures, let us consider a filter operation to be implemented in each of them. For *Von Neumann* architecture, multiplying two numbers requires at least three clock cycles — one to transfer the instruction and two more to transfer the two operands. It is assumed that the result remains in CPU register to accumulate with the future products of coefficients and signal values. Naturally, the entire computation is sequential and thus slow. The same filtering operation, when implemented in *Harvard* architecture, has the instruction in program memory and the operands in data memory. Thus, data memory bus is busier than the program memory bus. The introduction of instruction cache in *Super Harvard* architecture helps as follows. The program memory (as noted in Fig. 3.3(c)) holds both the instruction and one set of data (say, the filter coefficients). The other data set (signal values) are stored in the data memory. Most of the filtering algorithms involve loops with small body. When the program starts, it runs slowly, as the program instructions need to be copied to the instruction cache as well, apart from being fetched from the program memory. However, soon thereafter (typically after one iteration of the loop), the entire loop body stays in the instruction cache. Thus, in a single memory access cycle, the processor starts getting one filter coefficient from program memory and one signal value from data memory, which are operated upon based on the instruction available in the on-chip cache. Thus, it creates a high *memory access bandwidth.*

Another way of improving the Harvard architecture is to include a data cache which is dynamically loaded with data. DSPs like *TMS320C67xx* include both program and data cache, as shown in Fig. 3.4. In this processor, the *Level 1 (L1)* cache comprises 8 kilobyte of memory divided into 4 kilobyte of program cache and 4 kilobyte of data cache. The *Level 2 (L2)* comprises 256 kilobyte of memory divided into 192 kilobyte of mapped SRAM memory and 64 kilobyte dual cache memory, which can be configured as mapped memory, cache, or a combination of the two.

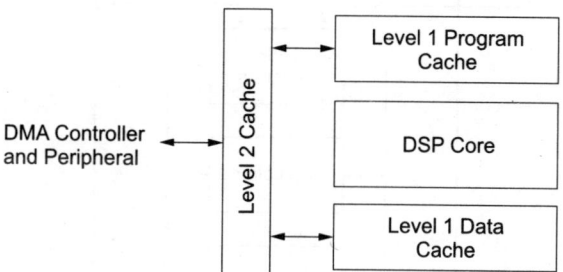

Fig. 3.4 Level 1 and Level 2 Cache in DSP.

Though cache memories improve the average performance of a DSP core, it makes the system a bit unpredictable. A cache miss can increase the execution time significantly. This is detrimental for hard real-time systems in which accurrate prediction of program execution time is a premier requirement. However, some modern cache management policies limit such effects by locking the cache to execute the time critical code fragments in a deterministic way.

3.2.2   Address Generation Units and Special Addresing Modes

The typical architecture of a DSP processor has been shown in Fig. 3.5. A marked feature of this architecture is one or more dedicated *address generation units*. Once the appropriate addressing registers have been configured, the address generation unit operates in the background (i.e., without using the main datapath of the processor) forming the addresses required for operand accesses in parallel with instruction execution. The address generation units are capable of generating addresses for various addressing modes. Several such addressing modes are generally supported, the most common being *register indirect with post-increment*. This is mainly used when a repetitive computation is performed on data stored sequentially in memory. *Modulo* addressing is also supported to simplify the use of *circular buffers*. Existence of several circular buffers helps in coding DSP algorithms in multiple stages, for example, IIR filtering. The address generators can also be configured to generate *bit-reversed* addresses into the circular buffers. This is necessary for FFT computation. Many implementations of the Fourier transforms require a reordering of either the input or the output data that corresponds to reversing the order of the bits in the array index. Carrying out the bit-reversal in software is very demanding and would result in using many CPU cycles that can be saved via hardware bit-reversal functionality. In the following passage we illustrate the concepts of circular buffer and bit-reversed addressing.

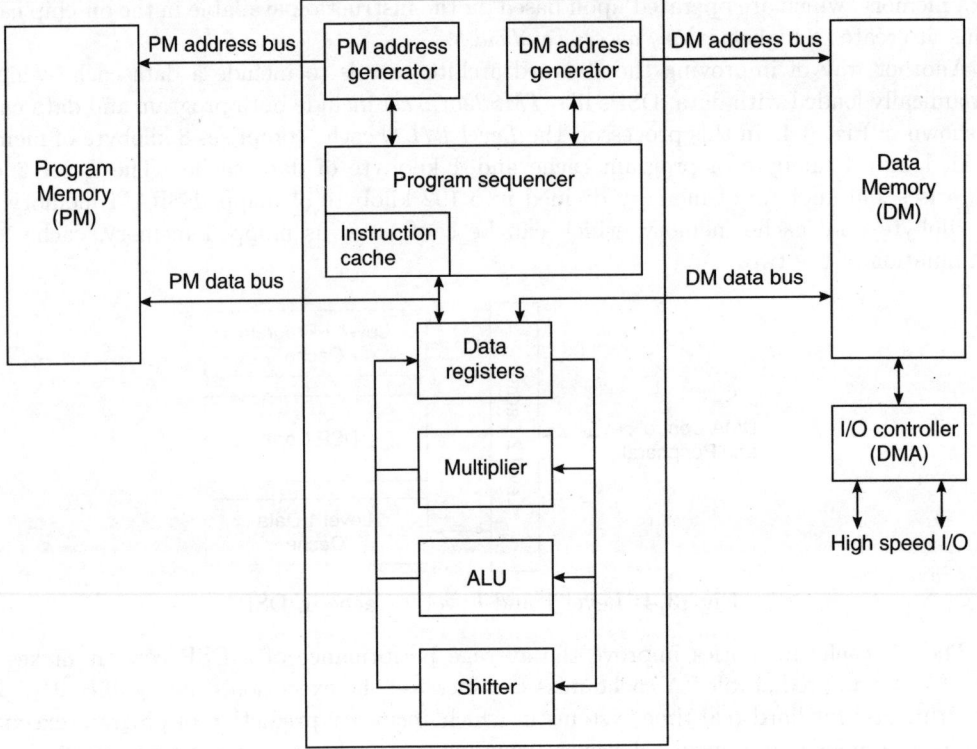

Fig. 3.5 Typical DSP architecture.

In the case of *linear buffers*, to store the new data element, each of the previous inputs are to be shifted downward, which is time consuming. The new data is written to the recently vacated slot at the top of the buffer. In *circular buffers*, a pointer is moved through the data, rather than moving the data itself. This requires that the pointer must be able to jump from the last location to the first, or vice-versa. Bit-reversed addressing provides the hardware support so that we do not need to check the pointer value in software after each entry to the buffer. Whereas for normal pointer arithmetic, carries propagate to the left, causing the values to grow until the highest number is reached, with bit-reversed addressing, the pointer arithmetic is done such that the carry bits propagate to the right. Carries from the rightmost bit are ignored, thus confining the pointer to a range determined solely by the value used to increment or decrement the pointer. For example, consider a buffer of size 8, an index register set to 4, and an initial pointer set to 0x100. Table 3.1 shows the adddress sequence generated by the bit-reversed addressing.

Table 3.1 Buffer address sequence

Address (hex)	3 LSBs	Comment
0x100	000	
0x104	000 + 100 = 100	
0x102	100 + 100 = 010	
0x106	010 + 100 = 110	Carry propagated to right
0x101	110 + 100 = 001	
0x105	001 + 100 = 101	
0x103	101 + 100 = 011	
0x107	011 + 100 = 111	
0x100	111 + 100 = 000	Carry falls off the right side

3.2.3   Direct Memory Access (DMA) Controllers

Direct Memory Access (DMA) is the hidden key for success of DSP systems. A DMA controller is a second processor working in parallel with DSP core and is dedicated to the transfer of data between two memory areas or between peripherals and memory. In the process, the DSP core becomes free to do other processing tasks. Figure 3.6 shows the typical position of a DMA controller within a generic DSP system. To have a better understanding of the situation, let us consider the following scenario. Suppose, in a system there is a 1 Msps (Mega symbols per second) ADC that produces 8-bit conversion data every microsecond. For an interrupt-driven data transfer to the DSP core, it generates about 1 million interrupts per second. Assuming each interrupt takes 30 CPU cycles, it requires about 5% of processor time, assuming the frequency of operation of the DSP core to be 600 MHz. The situation gets worsened further if there are multiple such ADCs integrated to the system. The DMA controller helps in this respect by taking the responsibility of transferring converted digital data samples to the memory, without bothering the DSP core.

There can be several modes of DMA data transfer, commonly known as *transfer configurations*. Two such configurations are as follows:

1. *Chained data transfer:* In this mode, the completion of one data transfer triggers a new transfer. This mode of transfer is useful mainly in applications requiring a continuous data stream as input. The situation has been shown in Fig. 3.7(a).

2. *Multidimensional data transfer:* This is particularly useful when the source and destination blocks are of different dimensions but the total number of memory locations in both

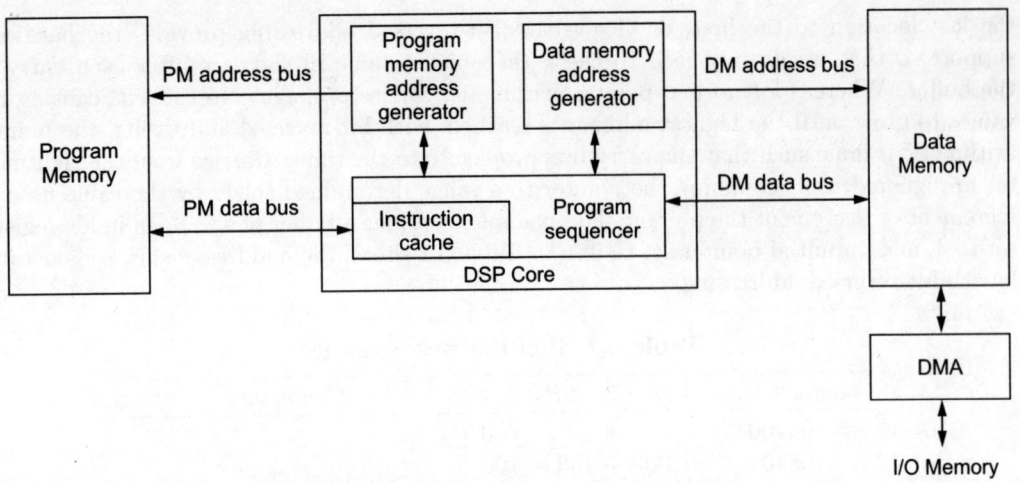

Fig. 3.6 Position of DMA controller within a generic DSP core.

the blocks are same. Figure 3.7(b) shows such a situation. Here, the source block is two dimensional from which a part of data is transferred to the destination which is a one dimensional array. The source and destination addresses, and the step value need to be set properly to accomplish the same. The step can be modified periodically to skip over the source memory locations. This type of data transfer is particularly useful for video applications.

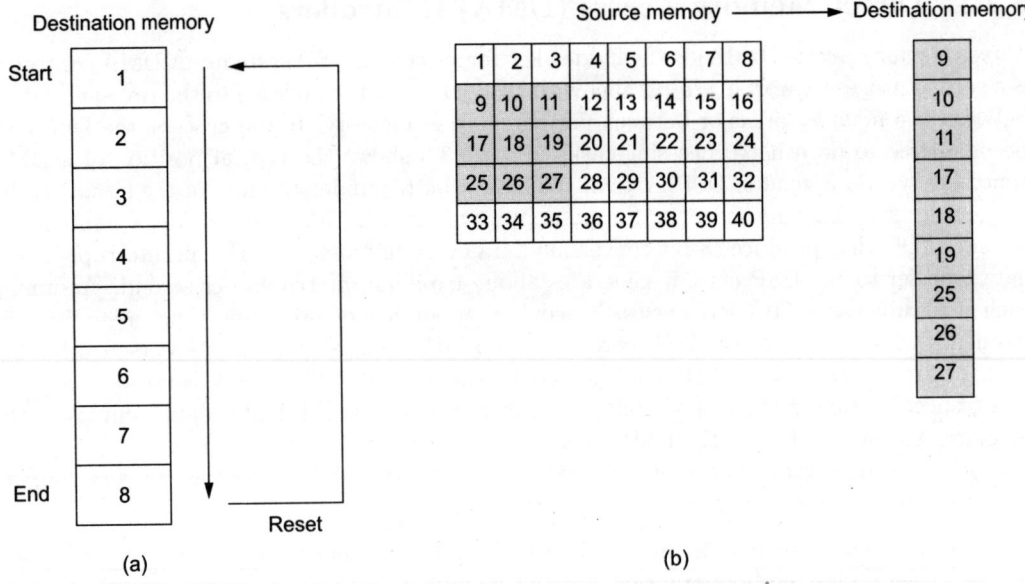

Fig. 3.7 DMA transfer configurations: (a) chained (b) multidimensional.

3.3   Fast Computation

Digital signal processors use a variety of techniques and architectures to achieve higher speed of computation. Such features include *Multiply-Accumulate* (MAC) units, instruction pipelining, and *Very Large Instruction Word* (VLIW).

3.3.1   Arithmetic Processing

Arithmetic processing is centred around MAC units. There can be one or more such units within a single DSP core. The basic arithmetic processing blocks are as follows:

1. *Registers:* The registers constitute the fastest memory in the DSP. Often there are large number of registers, and are wider than normal DSP word width. This helps in providing higher resolution during processing.

2. *Multipliers:* The multipliers can carry out single cycle multiplications. Extra wide accumulator registers are used to reduce round-off and truncation errors. Very often, an adder is also integrated into the multiplier.

3. *ALU:* The arithmetic logic unit carries out the arithmetic and logical functions.

4. *Shifters:* These are used to shift input values by one or more bits. The multi-bit shifters are also known as *barrel shifters* and are particularly used in floating point addition and subtraction operations.

3.3.2   Instruction Pipelining

This constitutes an important element in enhancing the performance of DSPs. The basic pipelining consists of three stages, as noted below:

1. *Fetch:* In this stage, the DSP generates the program fetch address and reads the opcode for the instruction.

2. *Decode:* This stage is responsible for sending the opcodes to the functional units, decoding of instructions, and reading the operands.

3. *Execute:* This stage carries out the actual execution of the instructions and then writes back the results to the registers.

As with any other pipelined architecture, branching and data dependency between instructions causes major difficulties in predicting the worst-case time requirement of a code sequence, which may be very important for real-time applications. Many DSP processors provide users techniques to control the pipeline. Such techniques include *time-stationary pipeline control*, *data-stationary control*, and *interlocked pipeline*.

3.3.3   Parallel Architectures

There are two levels of parallelism available in DSP processors—*instruction-level* and *data-level*. Instruction level parallelism leads to the *Very Large Instruction Word* (VLIW) architecture, while the data level parallelism leads to *Single Instruction Multiple Data* (SIMD) style of architecture.

1. *VLIW architecture:* In this case, a number of instructions are issued at the same time and are executed in parallel by a set of execution units. Such DSPs are also called *multi-issue* DSP. Instruction set is simple and regular in nature. Instruction scheduling is determined at compile-time, rather than run-time. Thus, the order of execution of instructions does not change, making the execution time predictable. The architecture is potentially scalable, more execution units can be added to allow a higher number of instructions to be executed in parallel. However, the memory bandwidth should be high, as a number of functional modules may access memory for operands. Due to the parallel activity, power requirement also goes up. The assembly level coding of an application becomes too difficult, the programmer has to depend on the usage of optimizing compilers.

2. *SIMD architecture:* These are based on data-level parallelism. A single instruction is issued and the operation is carried out on multiple data sets. SIMD and VLIW can also be coupled together within the same processor. However, for SIMD to be successful, there should be enough parallelism in operation. A strict sequential code cannot get the benefits of the features of SIMD computation. Often a *loop unrolling* (i.e., expanding the iterations of the loop) is performed to exploit the data-level parallelism.

3.4 Higher Accuracy

Since the DSPs are computationally intensive, it is very much essential to perform the arithmetic operations with higher level of accuracy. For this, several strategies have been adhered to in many of the DSPs. Some of the techniques are based upon the number representation formats while others are pure architectural enhancements. In the following, we enumerate some of these features.

1. *Data representation:* Based on their ability to handle real numbers, DSP can be divided into two categories — *fixed point* and *floating point*. Fixed-point DSPs usually represent each number by 16-bits. Thus, there are four common ways in which the $2^{16} = 65536$ different bit patterns can be used to represent a number. These are:

   - *unsigned integer*, storing 0 to 65535.
   - *signed integer* in two's complement form, storing numbers -32768 to 32767.
   - *unsigned fraction*, in which the 65536 levels are uniformly distributed between 0 and 1.
   - *signed fraction*, that can represent 65536 fractions in the range -1 to 1.

   On the other hand, floating-point DSPs typically use a minimum of 32 bits. Thus, the precision of floating point system is much higher than fixed point representation. This readily shows up in *signal-to-noise* ratio resulting from a repetitive computation. This happens as the number stored is actually a quantized value of the actual.

2. *Saturation arithmetic:* Many DSP processors handle the overflow problem in a different way as compared to general processors. A technique called *saturation arithmetic* is often utilized. In standard binary arithmetic, wrap-around is used for the values returned after an overflow or underflow. Thus, when 0111 and 1001 are added, the content becomes

10000. If the register is 4-bit long, the carry is lost, and the content becomes 0000. In saturating arithmetic, it tries to return a value which is as close as possible to the true result. That is, for overflow, it returns the largest possible value, while for underflow it returns the smallest possible one. For example, for the previous case, it returns 1111, instead of 0000. The scheme is particularly useful for video and audio applications. In such cases, the user will hardly understand the difference between the true value and the largest value that can be represented. This also ensures that no exceptions are generated due to overflow/underflow in real-time situation. Exceptions are difficult to accommodate in real-time systems.

3. *Large accumulator registers:* In a repetitive calculation, the storage of successive intermediary values results in accumulation of the quantization noise as well. DSPs often use an *extended precision accumulator* to handle this problem. This is a special purpose register having 2-3 times as many bits as other memory locations. This virtually eliminates round-off noise while the accumulation is in progress. Only when the final result is stored, error comes due to scaling of the number into 16 bits. The strategy is particularly used in connection with fixed point representation. For floating point case, quantization noise is so low that these techniques are generally not necessary. However, floating point DSPs are priced higher than their fixed-point counterparts.

3.5 Fast Execution Control

DSPs use quite a few techniques to execute control instructions faster. In the following, we look into two such aspects—*looping* and *interrupt response.*

1. *Zero Overhead Loops:* Any loop statement consists of the following parts—loop initialization, termination condition check, body of the loop, and updation of loop indices. A *fixed iteration loop* iterates its body for a fixed number of times. For example, the loop *for i = 1 to 100 do S_1* repeats the statement S_1 100 times. The initialization, condition check, and loop index updation are the loop overheads, as they require extra code over and above the code for the body part of the loop. For fixed iteration loops, DSPs provide hardware support so that no instruction is spent to code the overhead portions. Often a *loop* or *repeat* instruction is provided. This helps in implementing a *for* loop without expending any instruction cycles for updating and testing loop counter or branching back to the top of the loop-body. This feature is known as *zero-overhead* looping.

2. *Quick Interrupt Service:* Interrupts are served very quickly and in a deterministic way. When an interrupt occurs, DSP needs to save enough context information of the current task before starting the corresponding interrupt service routine (ISR). This context switching is done by the *interrupt dispatcher.* Many DSPs often contain more than one dispatchers. The user can select between them. Different dispatchers save different amount of context. A higher number of saved registers implies a longer context switching time. In many cases, alternate register sets are provided to minimize the context switching time.

3.6   C6000 Family of DSPs

The *C6000* family of digital signal processors from *Texas Instruments* are used in wide range of applications including audio and video processing, medical imaging, robotic control, etc. These DSPs have many architectural features that make them ideal for computation intensive real-time applications. The family includes fixed-point DSP cores *C64x* and *C64x+*, simple floating/fixed point DSP cores *C67x* and *C67x+*, and the newest floating/fixed point DSP cores *C674x* and *C66x*. They possess same high-level architectures and 100% upward compatibility. Of course, hardware enhancements and additional instructions are supported on the newer devices. The devices can be programmed very efficiently using *C/C++* languages. The high quality optimizing compilers ensure that the generated code achieves almost the same level of performance as assembly language code.

3.6.1   C6000 DSP Core

Figure 3.8 shows the high-level architecture of *C6000* DSP core. The components in the CPU are as follows:

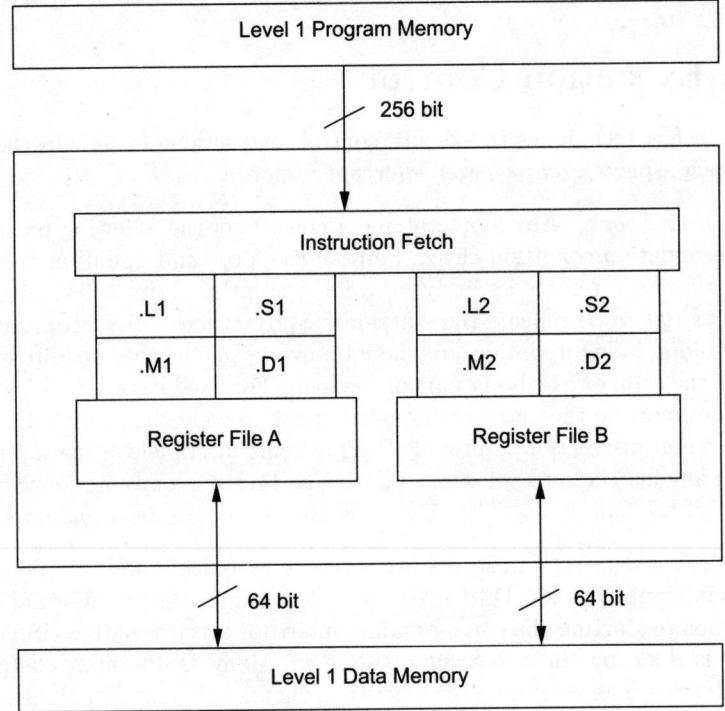

Fig. 3.8 C6000 DSP architecture.

1. *Eight parallel functional units.* Thus, upto eight instructions without data dependency can be executed in parallel. The data path elements are as follows:

   - *.D* (*.D*1 and *.D*2) handles data load and store operations.

- .S (.$S1$ and .$S2$) handles shift, branch and compare instructions.
- .M (.$M1$ and .$M2$) handles multiplication operations.
- .L (.$L1$ and .$L2$) handles logic and arithmetic operations.

2. *32, 32-bit registers.* Each side of functional units contains 32 such registers.

   - $A0$–$A31$ for A side (.$D1$, .$S1$, .$M1$, .$L1$)
   - $B0$–$B31$ for B side (.$D2$, .$S2$, .$M2$, .$L2$)

3. Program and data memory.

4. 256-bit internal program bus allowing 8, 32-bit instructions to be fetched from the program memory every cycle.

5. Two 64-bit internal data buses allowing .$D1$ and .$D2$ to fetch from data memory every cycle.

3.6.2   C6000 Pipeline

The standard *Fetch-Decode-Execute* pipeline of DSP processors has been further refined in C6000. Each of the three stages is divided into number of substages, and each substage finishing its processing in one CPU cycle. The stages are as follows:

1. *Fetch* has been divided into four substages:

   (a) PG: Program address generate—to update program counter register.
   (b) PS: Program address send—address sent to memory.
   (c) PW: Program access ready wait—waiting to get instruction from memory.
   (d) PR: Program fetch packet receive—one packet contains 8, 32-bit instructions.

2. *Decode* has been divided into two stages:

   (a) DP: Instruction dispatch—assign to the functional units.
   (b) DC: Instruction decode.

3. *Execute* may go upto 10 stages E1 through E10, depending upon the instruction to be executed.

Even with this finer level of pipelining, it sometimes beocomes necessary to halt the pipeline. Suppose that the current instruction is dependent on the result of the previous instruction, and that instruction takes more than one cycle for the result to be ready. For example, a *load* operation takes 5 clock cycles. Hence, after issuing a load instruction if the next instruction needs to refer to that value, the unit has to wait for 4 cycles. Same is the case with, say, multiplication that requires multiple clock cycles. Or, if a branch instruction is encountered, the branch target is known only after the instruction has entered the pipeline stage E1. Hence, the fetching of next instruction has to be delayed.

The cycles that the CPU has to wait are called *delay slots* of the instructions. For example, a load operation has four delay slots, etc. These delay slots can cause severe performance degradation, particularly within a loop body. The delay slots are attempted to be filled using *software pipelining* and *SPLOOP buffer*.

3.6.3   Software Pipelining

This is a technique to schedule instructions from a loop so that multiple iterations of the loop can be executed in parallel and the pipelining can be exploited effectively. For example, consider the task of adding 15 values stored in memory. In a traditional processor, it will employ a loop of 15 iterations, where each iteration does the following:

- Load next value
- Add

Table 3.2 Traditional scheduling vs. Software pipelining

Traditional Scheduling			Software Pipelining			
Clock	.D1	.L1	Clock	.D1	.L1	
1	Load1		1	Load1		
2			2	Load2		
3			3	Load3		Prolog
4			4	Load4		
5			1	Load5		
6		Add1	6	Load6	Add1	
7	Load2		7	Load7	Add2	
8			8	Load8	Add3	
9			9	Load9	Add4	
10			10	Load10	Add5	
11			11	Load11	Add6	Kernel
12		Add2	12	Load12	Add7	
13	Load3		13	Load13	Add8	
14			14	Load14	Add9	
15			15	Load15	Add10	
16			16		Add11	
17			17		Add12	
18		Add3	18		Add13	Epilog
19	Load4		19		Add14	
20			20		Add15	

However, loading from memory takes 5 clock cycles, addition takes 1 clock cycle. Thus, each iteration takes 6 clock cycles, making a total of 90 clock cycles for 15 iterations. The situation has been shown in the first column of Table 3.2. A better way of scheduling instructions across the iterations has been shown in the second column of Table 3.2. As it can be observed from the figure, after executing *load1*, *load2* is executed. After that, at every clock one more *load* is executed. The first load produces its result after 5 clock cycles, hence, in the 6th clock cycle, first addition could be done. However, in successive cycles one load and one add are performed parallely. This continues upto clock cycle 15. In clock cycles 16 through 20, the remaining additions are done. The clock cycles 6 to 15, in which parallel execution takes place, constitute the *kernel* of the software pipelining. Clock cycles 1 to 5 make the *prolog* while the clock cycles 16 through 20 form the *epilog* of the pipeline. However, it should be noted that such a software pipelining is difficult to be carried out manually. The C compiler of C6000 can be instructed to perform such a pipeline.

3.6.4 Software Pipelined Loop (SPLOOP) Buffer

The software pipelining discussed above has two major drawbacks. Firstly, it is the kernel portion that actually carries out the task. However, code must be added to perform the prolog and the epilog portions. This increases the code size. Secondly, at any point of execution, multiple iterations of the loop are active. Hence, there may exist a large number of live variables. In case of occurrence of an interrupt, all these need to be saved. This may affect performance and increase the code size. The $C64x+$ CPU has added a software pipeline buffer (shown in Fig. 3.9). The buffer stores a single scheduled iteration of the loop, along with some *framing* SPLOOP control instructions. The control instructions mark the beginning and end of a SPLOOP block, number of iterations (fixed/variable) etc. The loop's instructions and pipeline characteristics are built up in the buffer. This removes the necessity of prolog and epilog code. It also reduces program fetches and thus the power requirement. Moreover, since the DSP core does not make out the difference about whether the instruction is coming from SPLOOP buffer or memory, the loops in the SPLOOP buffer are fully interruptible, no special care is required to maintain the state. However, a SPLOOP buffer cannot be used to handle loops that exceed 14 execute packets. An execute packet can contain upto 8 instructions that can be executed in parallel. It is difficult to handle complex loops in this type of environment. Hence, it is recommended that the loops should be simple and short in length.

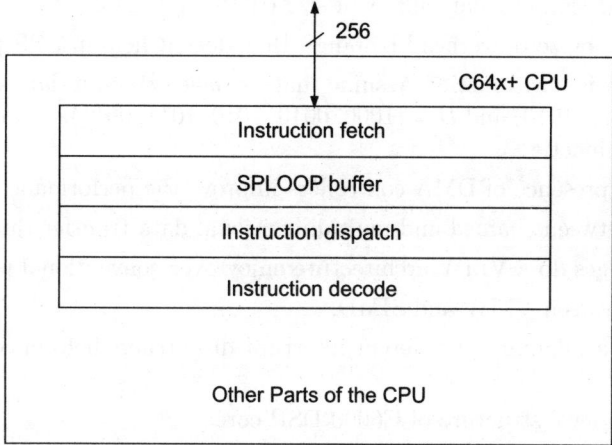

Fig. 3.9 Part of C64x+ CPU showing SPLOOP buffer.

3.7 Conclusion

In this chapter we have seen an overview of the special features of digital signal processors. Availability of such features make DSP ideal for embedded tasks involving signal processing. Whereas, in the next chapter we will discuss another often used platform for embedded system design—*Field Programmable Gate Array* (FPGA).

Exercises

3.1 What are the different hardware platforms available for embedded system realization? Perform a comparison between these from different design aspects.

3.2 What is a digital signal processor and how does its characteristic differ from a general-purpose processor?

3.3 What is the typical structure of a MAC unit of DSP processor? Why do you think that such a module in the datapath of a DSP processor is essential?

3.4 A polynomial $P(x) = a_0 + a_1 x + a_2 x^2 + \ldots a_{n-1} x^{n-1}$ can be viewed as $P(x) = a_0 + x(a_1 + x(a_2 + x(a_3 + \ldots + x(a_{n-1})) \ldots)$. Show that a datapath using MAC structure can be used to evlaute it.

3.5 Distinguish between Von Neumann, Harvard and Super Harvard architectures.

3.6 Suppose we have to compute $Y = \sum_{i=0}^{n-1} a_i x_i$. Assume that the total number of memory accesses for the program code is 20 words, each a_i and x_i values occupy one word each, the result Y is stored in a register. If memory access takes 10 ms and multiply-add takes 10 ns, compute the time needed to execute the program in Von Neumann, Harvard, and Super Harvard architectures. For Super Harvard case, assume that the program is available in cache 80% of time and one cache access takes 2 ns. Assume the value of n to be 100.

3.7 How does modulo arithmetic differ from auto-increment/decrement mode of addressing?

3.8 How does bit-reversed addressing help in implementing circular buffer? Use bit-reversed addressing to design circular buffers of size (i) 16, and (ii) 4.

3.9 What is meant by zero-overhead looping? How does it help in DSP program execution?

3.10 What is saturation arithmetic? Assume that we have two 4-bit data streams: $A = (0111, 1000, 0000, 1111, 1010)$ and $B = (1000, 0010, 1010, 1011, 0011)$. Compute $\sum_i A_i B_i$ using saturation arithmetic.

3.11 How does the presence of DMA controllers improve the performance of a DSP system?

3.12 Distinguish between chained and multidimensional data transfer through DMA.

3.13 What advantages do a VLIW architecture enjoy over conventional processors?

3.14 Distinguish between VLIW and SIMD.

3.15 How does the availability of different interrupt dispatchers help in controlling the interrupt latency?

3.16 Show the high level structure of C6000 DSP core.

3.17 What is extension precision accumulator and how does it help to improves signal-to-noise ratio in DSP calculations?

3.18 Show a software pipelined execution of a task that adds twenty numbers stored in memory.

CHAPTER 4

Field Programmable Gate Arrays

The two target architectures for embedded systems—microcontrollers and digital signal processors, studied in the previous two chapters provide software based solutions. However, where the performance becomes a very important criteria, we have to look for alternatives. Such alternatives include *Application Specific Integrated Circuit* (ASIC), *Field Programmable Gate Array* (FPGA) etc. While ASIC design can be targeted to any application, the NRE cost is very high for them. *Field Programmable Gate Arrays* (FPGAs) have been introduced as an alternative to custom ICs for implementing circuits in hardware, at the same time providing flexibility of reprogrammability to the user. A comparative analysis between general processor (including DSP), FPGA and ASIC is presented in Fig. 4.1.

Performance	Non-recurring engineering cost	Unit cost	Time-to-Market
ASIC	ASIC	FPGA	ASIC
FPGA	FPGA	General processor	FPGA
General processor	General processor	ASIC	General processor

Fig. 4.1 Comparison between target architectures.

A complete discussion on ASIC design is beyond the scope of this book. In this chapter we will have an overview of commonly available FPGAs and their variations.

4.1 Field Programmable Devices

A *field programmable device* refers to any kind of integrated circuit in which the chip can be configured by the end user to realize any design. The advantages provided by such a device are instant manufacturing turnaround, low start-up costs, low financial risk and ease of design changes (reprogrammability). However, there exists large number of such devices. In the following we note some of the important categories.

- PLA (*Programmable Logic Array*). It contains two levels of logic, an AND-plane and an OR-plane, where both levels are programmable.

- PAL (*Programmable Array Logic*). It has a programmable AND-plane followed by a fixed OR-plane.

- SPLD (*Simple Programmable Logic Device*) and CPLD (*Complex Programmable Logic Device*). SPLD refers to any simple programmable logic devices like PLA/PAL. CPLD consists of an arrangement of multiple SPLD like blocks on a single chip.

- FPGA (*Field Programmable Gate Array*). It is a field programmable device having a general structure that allows very high logic capacity. While CPLDs normally have logic resources with a wide number of inputs, FPGAs possess narrower logic blocks. The ratio of flip-flops to logic resources is also higher in FPGAs, compared to CPLDs.

Out of all these field programmable devices, FPGAs support very high logic capacity. Thus, it has been instrumental for a major shift in the way digital systems are designed. This is particularly suitable for embedded system design, as to have application specific hardware, we do not need to have a custom IC, a reprogrammable FPGA chip serves the purpose at a much lower cost with added ease of design modification.

A generic FPGA structure has been shown in Fig. 4.2. It consists of regularly arranged array of logic blocks. The logic blocks can be programmed to realize different Boolean functions. The blocks often contain one or more flip-flop(s) in them. Between the rows and columns of logic blocks, interconnect links run (in parallel) that can be used to transfer signal values from one logic block to some other one. The programmable switches present at the junction of interconnect rows and columns can be utilized for this purpose. I/O blocks run at the boundary of the chip to accept input from the environment, and producing output to the environment.

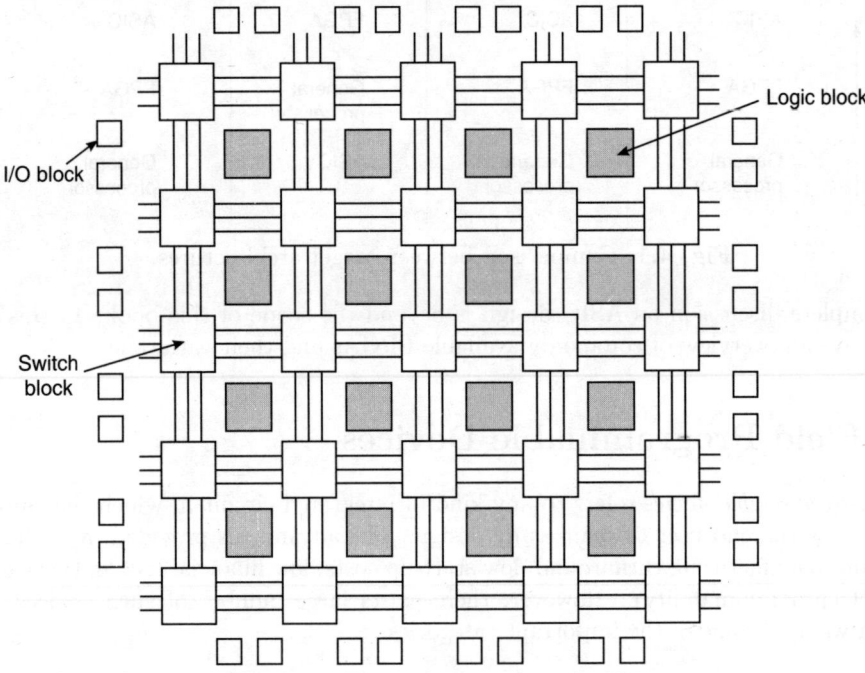

Fig. 4.2 Generic FPGA architecture.

4.2   Programmability of FPGA

User programmability of CPLDs and FPGAs is achieved via user-programmable switch technologies. For CPLDs, floating-gate transistors are used like EPROM or EEPROM. On the
other hand, FPGAs normally use SRAM (static RAM) or antifuse technology. Properties of
the switches, such as, *size, on-resistance*, and *capacitance* dictate trade-offs in architecture. In
SRAM based FPGAs, there is an SRAM bit corresponding to each of the programmable points
within the device. When the device is powered-on or reset, it reads a configuration program
from an off-chip memory and loads it into on-chip SRAM. The configuration program defines
the logic function realized by individual logic blocks and interconnections. Figure 4.3 shows
an example in which SRAM bits are controlling the switches to route output of one logic block
to the input of another. Devices using SRAM based switching can be reprogrammed easily
by just changing the configuration program. FPGAs belonging to *Xilinx, Plassey, Algotronix,
Concurrent Logic, Toshiba*, etc. are SRAM-based. It provides fast reprogrammability at the
cost of large area (at least five transistors for cell and one for switch).

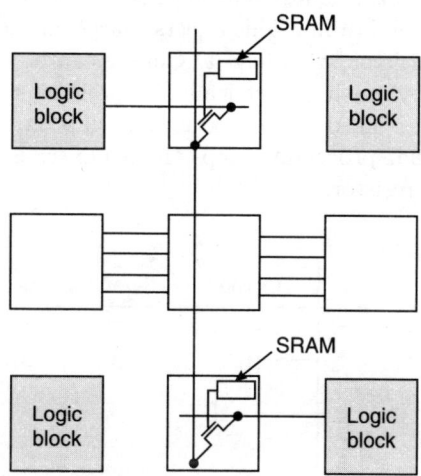

Fig. 4.3 SRAM controlled switching.

Another widely used programmable switch in FPGA is the *antifuse*. Antifuses are originally open-circuit, offering very high resistance. However, on programming (applying a 11-20V
across terminals), the resistance becomes very low, thus establishing electric connections. Antifuses can be made very small using modified CMOS technology, thus offering very high device
density, compared to SRAM. However, once programmed, they cannot be reused. Thus, the device is one-time programmable. An example of antifuse used in Actel FPGAs has been shown in
Fig. 4.4. The structure is commonly known as PLICE (Programmable Low-Impedence Circuit
Element). The figure shows an antifuse positioned between two interconnect wires and physically consists of three sandwitched layers—the top and bottom layers are conductors, while
the middle layer is an insulator. In the unprogrammed state, the insulator isolates the top
and the bottom layers, however, on programming, the insulator changes into a low-resistance
link. PLICE uses Poly-Si and n^+ diffusion as conductors and ONO (silicon diOxide–silicon
Nitride–silicon diOxide) as an insulator. The advantages include small size (little more than

the cross-section of two metal wires) and low series resistance. It has disadvantages, such as large size of programming transistors, need of isolation transistors, and one-time programmability. FPGAs from *Actel, Quicklogic, Crosspoint,* etc. support antifuses.

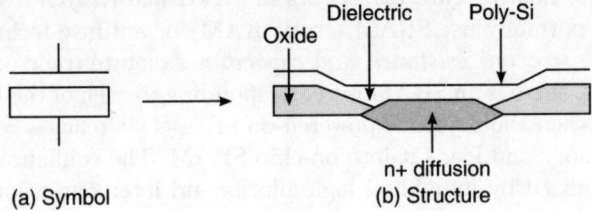

Fig. 4.4 Antifuse structure.

FPGA devices from *Altera, Plus Logic, AMD,* etc. use floating gate programming technology. While *Altera* and *Plus Logic* use ultraviolet erasable EPROM, *AMD* uses electrically erasable EEPROM. Figure 4.5 shows a programmable switch based on floating gate. It contains a control gate and a floating gate. The transistor can be disabled by applying a high voltage between control gate and drain. This injects charge on the floating gate, increasing the threshold voltage of the transistor, disabling it. Charge can be removed by exposing floating gate to ultraviolet light or by erasing electrically. It provides reprogrammability and unlike SRAM, no external memory is needed to program the chip on power-up. However, EPROM technology requires additional processing steps, high ON resistance and high static power consumption due to pull-up resistor.

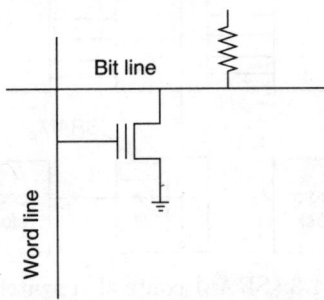

Fig. 4.5 Floating gate programming technology.

A comparison of switching techniques has been shown in Table 4.1.

Table 4.1 Comparison of programming techniques

Name	Whether reprogrammable	Whether volatile	Technology
Fuse	No	No	Bipolar
EPROM	Yes, out of circuit	No	UVCMOS
EEPROM	Yes, in circuit	No	EECMOS
SRAM	Yes, in circuit	Yes	CMOS
Antifuse	No	No	CMOS+

4.3 FPGA Logic Block Variations

There are wide variations in the logic block structure of FPGAs available from different vendors. They vary in number of inputs and outputs, amount of area consumed, complexity of logic functions that they can realize, total number of transistors needed, and so on. The logic blocks can broadly be classified into the following two categories:

1. *Fine-grain:* The block contains a few transistors that can be interconnected via programming. *Crosspoint* FPGA uses a single transistor pair for each boolean variable in the logic block. An example implementation of the function $f = ab + \bar{c}$ has been shown in Fig. 4.6.

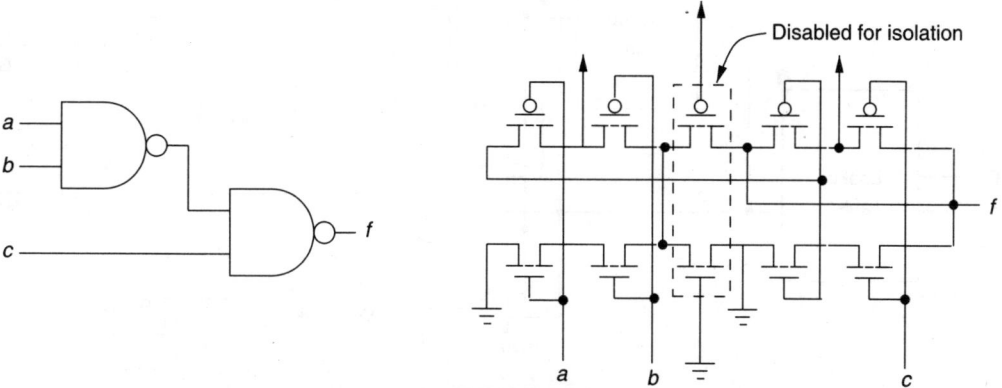

Fig. 4.6 Crosspoint realization of function $f = ab + \bar{c}$

2. *Coarse-grain:* The block contains good amount of logic in it. The complexity of the block varies a lot. For example, in the following paragraph, we will look into two different logic blocks with wide variation in complexity — *XC4000* from *Xilinx* and *Act1* from *Actel*.

Figure 4.7 shows the logic block of Xilinx XC4000 FPGA. It is commonly called a *Configurable Logic Block* (CLB). It is based on *Look-Up Tables* (LUTs). An LUT is a one-bit SRAM-based memory array. The address lines are the inputs to the LUT and the one bit output from the memory is the LUT output. Thus, an LUT with k number of inputs need a memory of size $2^k \times 1$-bits. By programming this memory suitably, any of the 2^{2^k} Boolean functions can be realized by the LUT. The XC4000 CLB contains two 4-input LUTs and another 3-input LUT that can be used in conjunction with these two to realize a wide range of logic functions upto nine inputs, two separate functions of four-inputs or other functions as depicted by the structure. Each CLB also contains two flip-flops. The advanced versions of Xilinx FPGAs possess CLBs with higher capabilities.

On the other hand, Actel FPGA logic blocks are built around 2–to–1 multiplexers. One such block (called ACT1) has been shown in Fig. 4.8. By restricting the inputs to some constant values, or shorting two or more of them, different logic functions can be implemented. Though the set of realizable logic functions is not as large as in Xilinx CLB, ACT1 is a very simple block, and thus has very low overhead. It may be noted that the advanced versions of Actel FPGAs also have improved logic blocks.

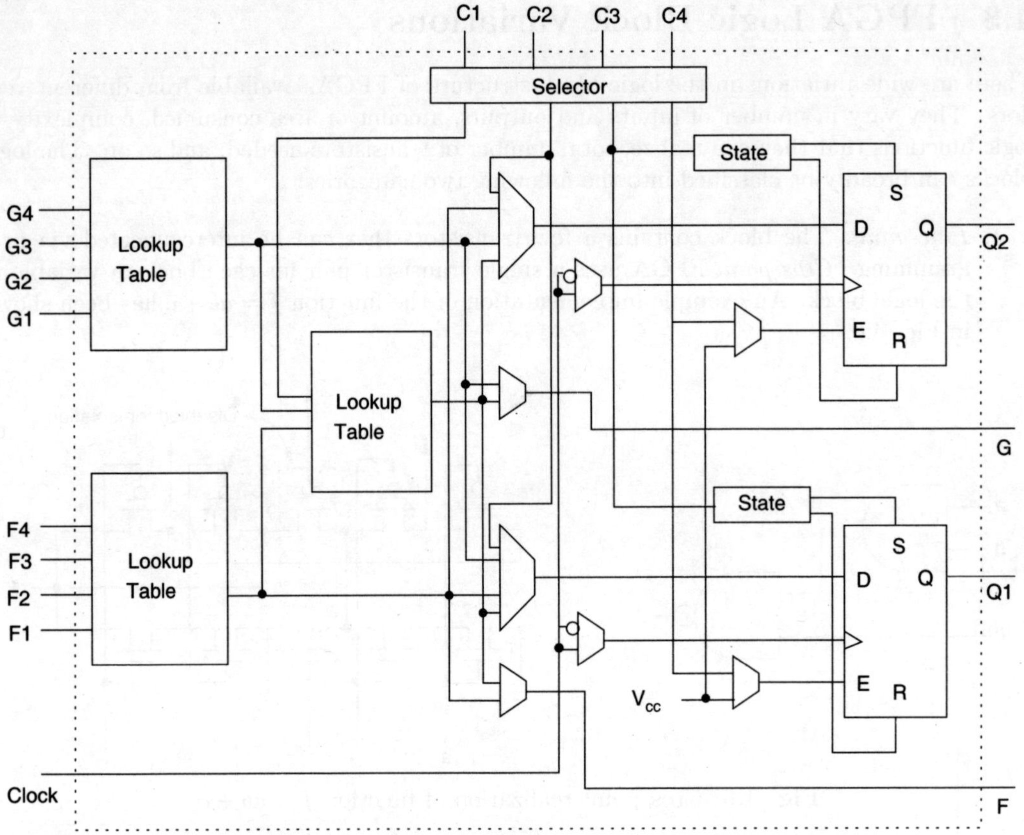

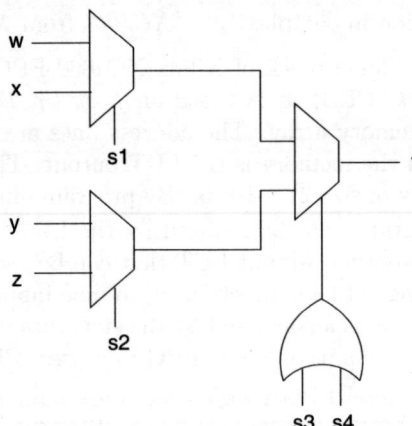

Fig. 4.7 Xilinx XC4000 configurable logic block (CLB).

Fig. 4.8 Actel ACT1 logic block.

There exists a definite trade-off between size of logic block and system performance as noted next.

- Since a large logic block can implement more logic within a single block, it requires lesser number of logic blocks to realize a given functionality on the FPGA. On the other hand, a large logic block consumes more space of FPGA.

- In case of look-up table based FPGAs, a 4-input look-up table gives best result in terms of logic synthesized and area consumed.

- Granularity of logic block has influence on performance of an FPGA. A higher granularity level results in lesser delay between system input and output. This is expected since lesser number of switched routing resources are needed in the design, since a single block realizes a bigger chunk of logic. On the other hand, with the increase of granularity level, average fanout increases, number of switches also increases as each block has more pins. Also, the length of wires increases with increase in size of logic block.

4.4 FPGA Design Flow

Vendors of individual FPGAs come up with an integrated CAD environment that is to be used for system development on their FPGA. A typical design flow has been shown in Fig. 4.9. It consists of the following components:

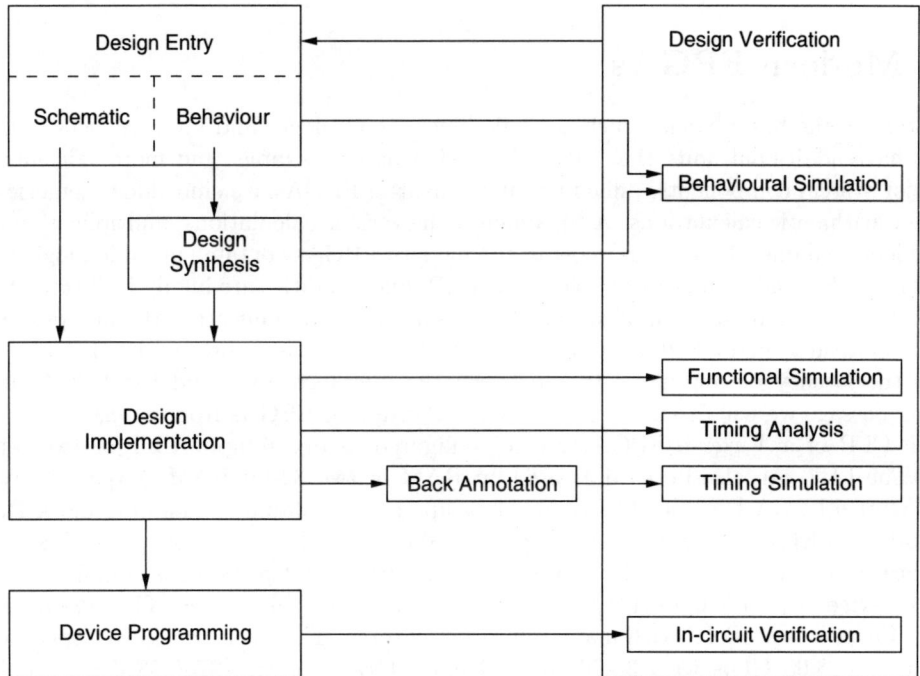

Fig. 4.9 FPGA design flow

1. *Design entry:* The design entry may be done either via schematic entry or via behavioural description. The schematic entry procedure uses a graphical interface to allow the designer to enter the design in terms of library modules and their interconnections. The behavioural entry can be made via hardware description languages, such as VHDL, Verilog etc. The behavioural description goes through the stage of design synthesis to generate a netlist of library components to realize the functionality. Schematic is already a netlist, and hence, does not require this stage.

2. *Behavioural simulation:* This is necessary to establish the correct functionality of the system at schematic/behavioural level.

3. *Design implementation:* This stage implements the generated netlist for the target device. It often involves translation between the formats. A functional simulation is performed to check the functional correctness of the translation. Next, detailed place and routing is carried out to get the implementation of the design onto the target device. Timing analysis reports the net delays. These delay values are back annotated onto the design and the design is simulated using a timing simulation. This helps in identifying whether there exists any timing violations (such as, set-up and hold time violations) in the implementation.

4. *Device programming:* This stage generates the final configuration program for the device. The configuration program may be downloaded onto a memory to be read during power up (for SRAM FPGAs), or may be used to generate antifuse/EPROM burning pattern. Many FPGAs contain in-circuit verification module that can be utilized to check the correctness of the downloaded design.

4.5   Modern FPGAs

In addition to the basic blocks (such as, logic blocks, I/O blocks and interconnects), modern FPGAs have additional units that make the design process simpler and more efficient. The two major system components, difficult to implement in FPGAs are embedded memories and blocks for arithmetic calculations. Also, amongst the various calculations, multiplication is the most widely used one. As a result, most of the modern FPGAs contain embedded logic blocks for multiplication and memories to hold data. DSP functionalities are highly facilitated by the availability of these. In many applications, FPGAs need to communicate with microprocessors. This has motivated many FPGA vendors to embed soft processor cores within FPGAs. This reduces the latency of communication between the microprocessor and the FPGA. In the following passage we will discuss about some such advanced FPGAs from *Xilinx*.

Each CLB of a *Virtex-6* FPGA can be configured as one 6-input LUT or two 5-input LUTs. The LUT can also be used as a 64-bit RAM or two 32-bit RAMs. Apart from this, every *Virtex-6* FPGA has 156-1064 (depending upon the subfamily) dual port block RAMs, each storing 36 Kbits. They also possess many dedicated, full-custom, low-power DSP slices. Each slice contains 25, 18-bit, 2's complement multiplier and a 48-bit accumulator. Each *Virtex-6* device has a 17-channel, 10-bit ADC and 8-72 Gbps transceiver. The next advanced version, *Virtex-7* is a 3D IC with many improved features. The peak transceiver speed varies between 12.5-28.05 Gbps with 36-96 transceivers. It can perform 2756-5314 giga multiply accumulates (GMACs) and contains 46.5-85 Mb block RAM, PCI express bus interface, and upto 1200 I/O pins.

4.6   Conclusion

In this chapter, we have looked into another very commonly used target in embedded systems – FPGA. It has the potential to give application specific circuitry at a much lower cost compared to ASIC. We have reviewed two architectures here. However, it should be mentioned that there exists a good number of other FPGAs available from different vendors.

Exercises

4.1 What is FPGA and how does it differ from ASIC?

4.2 Distinguish between PLA, PAL, SPLD, CPLD, and FPGA. Give examples of at least one device from each category.

4.3 Distinguish between programming technologies based on SRAM, antifuse, and floating gate. Give examples of at least one FPGA for each category.

4.4 Enumerate the trade-offs between fine-grain and coarse-grain logic blocks.

4.5 What is the total number of combinational functions that can be realized using either of the lookup tables (LUT) with inputs $G1$ to $G4$ and $F1$ to $F4$ in the CLB of Xilinx FPGA? Also, if both of them are utilized to implement a single function using the LUT and $C1$, what is the total number different functions realizable?

4.6 Write a program (in C/C++) to find out the number of non-equivalent functions of various inputs realizable using the ACT1 block. Two functions are equivalent if one of them can be realized from the other using only a permutation of inputs.

4.7 Enumerate the FPGA design flow.

4.8 Perform a detailed study on the features of *Virtex-6* FPGAs.

4.9 Perform a detailed study on the features of *Virtex-7* FPGAs.

CHAPTER 5

Interfacing

Interfacing plays a very important role in connecting processors to the peripheral devices, or even in communicating between the processors. As there are wide variety of processors available, there also exist many peripheral devices to be connected to them to complete an embedded application. The interfacing requirements of these devices vary a lot. However, to keep the overall design simple, it is very much necessary to make the devices and processors compatible with each other through a set of design parameters to enable communication between them. The common ways of such interfacing are:

- Synchronous interfaces, like *SPI, I^2C*, etc.
- *RS232C, RS422* based serial interface, *USB*.
- Network interfaces, such as *RS485, CAN, Ethernet*, etc.

In this chapter, we will have a look at some of these interfaces.

5.1 Serial Peripheral Interface (SPI)

Serial Peripheral Interface has been developed by *Motorola* to provide a simple, low-cost interface between microcontrollers and peripheral chips. The interface is also known as a *four-wire interface* as it uses four main signals to enable the data transfer. These signals, as shown in Fig. 5.1, are:

1. *MOSI:* Master Out Slave In
2. *MISO:* Master In Slave Out
3. *SCLK:* Serial Clock
4. *CS:* Chip Select

The *SPI* interface can be used for interfacing memory, ADC, DAC, real-time clock, LCD drivers, sensors, audio chips, and even other processors. The following are a few examples of such devices.

- *Temperature sensor:* LM74 having 12-bit plus sign temperature resolution (0.0625°C per LSB) with a temperature range of −55°C to +150°C.
- *Pressure sensor:* SCP1000 with 17-bit resolution. Under ideal conditions it can detect the pressure difference within a 9 cm column of air.
- *Analog-to-Digital converter:* LTC2452 is an ultra-tiny, differential, 16-bit delta-sigma ADC.

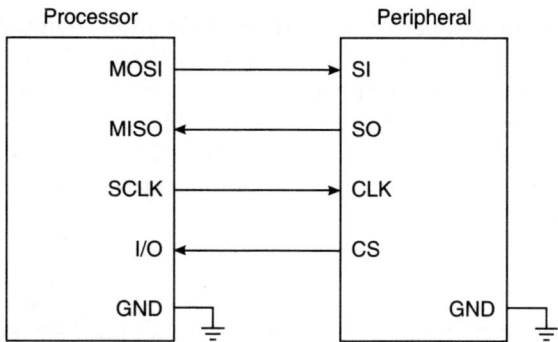

Fig. 5.1 The SPI interface.

- *Touch screen:* SX8652 is a very low power, high reliability controller for resistive touch screens. Supports wide input supply range of 1.65 V to 3.7 V.

- *EEPROM:* Microchip 25XXX serial EEPROM, with densities from 128 bits to 512 Kbits, 3-20MHz bus speed, low power operation, built-in write protection.

- *Real time clock:* DS3234 is a low-cost extremely accurate real-time clock with integrated temperature-compensated crystal oscillator and crystal. It counts seconds, minutes, hours, day, date, month, year with leap year compensation, valid upto 2099.

- *Memory card:* MMC and SD cards.

The most important feature of *SPI*, as compared to a standard serial port is that *SPI* is a synchronous protocol in which all transmissions are referenced to a common clock, generated by the master. The receiving peripheral (slave) utilizes the clock to synchronize the transmission.

Both the master and slave contain a serial shift register. The contents of these shift registers are exchanged to facilitate data transfer between the master and slave. The processor (master) initiates the transfer by writing a byte to its SPI shift register. As the register transmits the byte to the slave on the *MOSI* signal line, the slave transfers the content of its shift register back to the master on the *MISO* signal line. Thus, the contents of the two registers are exchanged (Fig. 5.2). Both a write and a read operation are performed with the slave simultaneously. If only the write operation is desired by the master, it simply ignores the byte it receives from the slave. On the other hand, if only a read operation is required by

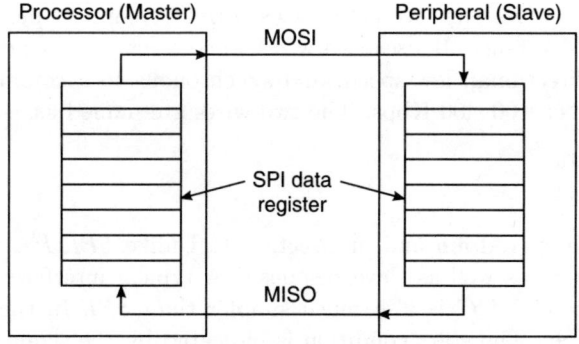

Fig. 5.2 Data transfer through SPI interface.

the master, it must transfer a dummy byte to the slave in order to initiate a slave transmission. Transmission often consists of 8-bit words, however, other word sizes are also common (for example, 16-bit words for touch-screen controllers and audio codecs, 12-bit words for many DACs and ADCs). Multibyte transmission is also possible. More than one peripheral chip can be connected to the same master through *SPI* interface. The slaves are selected by the master by asserting the corresponding chip-select input. A multi-peripheral interface has been shown in Fig. 5.3.

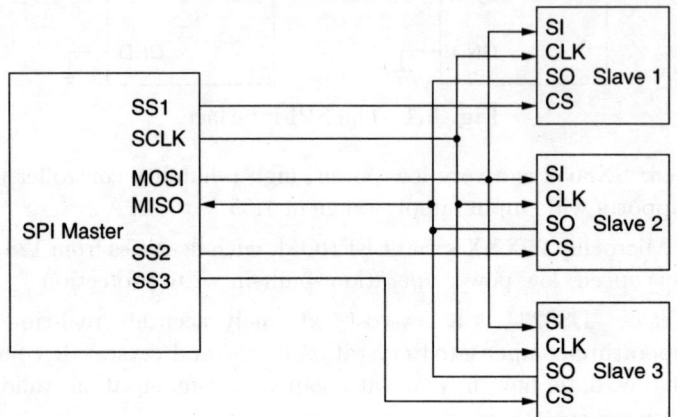

Fig. 5.3 Connecting multiple slave devices to a master.

Conventionally, apart from setting the clock frequency, the master also configures the polarity and phase of the clock signal with respect to data. This is done by setting the *CPOL* and *CPHA*. These often constitute two bits of the SPI control register. With $CPOL = 0$, the base value of the clock is taken as zero. Now, if $CPHA = 0$, data are captured on the rising edge of the clock and transmitted at the falling edge. With $CPHA = 1$, data are captured on the falling edge of the clock signal and transmitted on the rising edge. With $CPOL = 1$, the base value of clock signal is taken as 1. The situation has been shown in Fig. 5.4.

5.2   Inter-Integrated Circuit (*IIC*, *I²C*)

I^2C bus is a very cheap, yet effective network used to interconnect peripheral devices within small-scale embedded systems. It uses two wires to connect multiple devices in a multidrop bus. The bus is bi-directional, low-speed, and synchronous to a common clock. Achievable data rate varies between 100–400 Kbps. The two wires are named as,

- *SDA:* Serial Data
- *SCL:* Serial Clock

Both the lines are open-drain and bi-directional. Unlike *SPI*, I^2C uses same signal line for master transmission, as well as slave response. A typical interface is shown in Fig. 5.5. The mode of operation of I^2C is also much simpler than *SPI*. In the idle condition, both *SDA* and *SCL* are high. The start condition is indicated by the sequence of *SDA* going low followed by *SCL* going low. Now *SDA* transitions for the first valid data bit. For each bit that

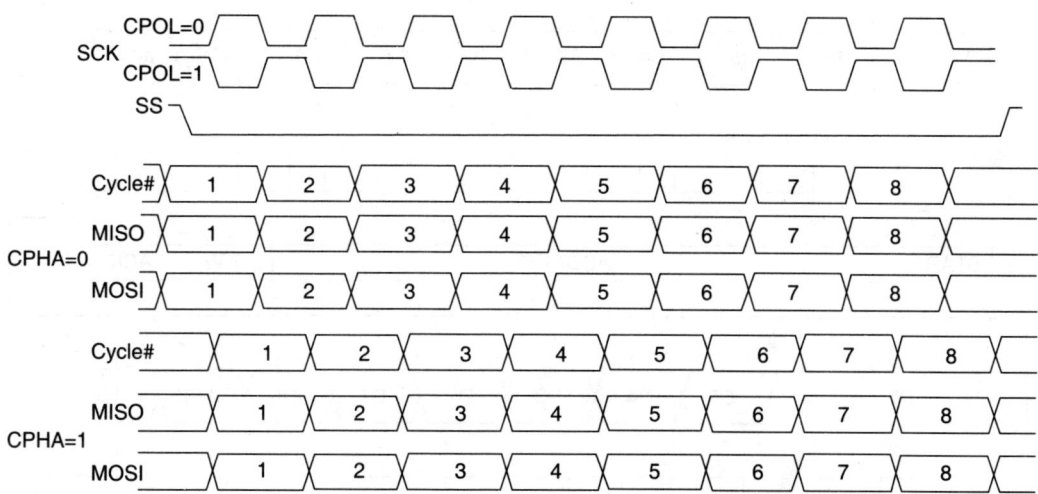

Fig. 5.4 Timing diagram for SPI communication.

is transmitted, it must become valid on *SDA* while *SCL* is low. The rising edge of *SCL* is used to sample the bit. The bit must remain valid till *SCL* goes low once more. Now, *SDA* transitions to the next bit, before *SCL* goes high once more. The stop condition is indicated by *SCL* returning to high followed by *SDA*.

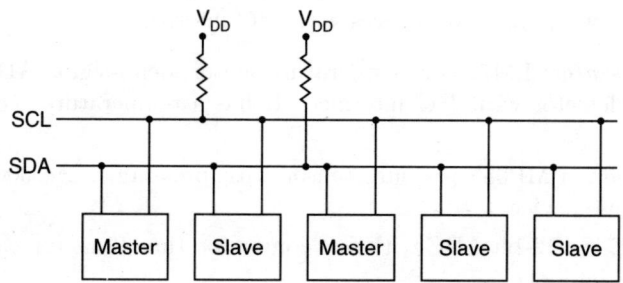

Fig. 5.5 Data transfer through *I²C* interface.

Any number of bytes can be transmitted in an *I²C* packet. If the receiver is unable to accept any more bytes, it can abort transmission by holding *SCL* low. In that case, transmitter waits till *SCL* is released by the receiver.

To facilitate acknowledgement, after transmitting the 8-th bit, master releases the *SDA* and gives an additional clock pulse on *SCL*. This triggers the receiver to acknowledge the byte by pulling *SDA* low.

Since there can be multiple devices on the same bus, each device has a unique 7-bit address. The first byte transmitted is the address byte along with a direction bit. The direction bit '0' indicates a *write* operation in which the slave will receive data, whereas the direction bit '1' indicates a *read* operation in which the slave will send data. Figure 5.6 shows the timing diagram of *I²C* transfer.

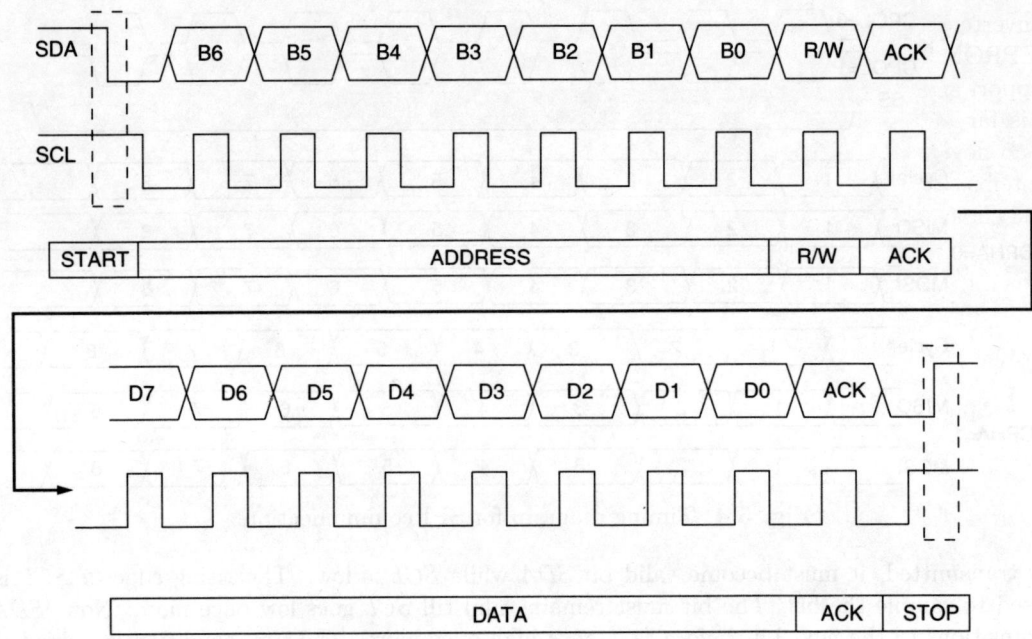

Fig. 5.6 Timing diagram for I^2C transfer.

Following are a few examples of devices with I^2C interfaces.

- *Temperature sensor:* LM75 is a temperature sensor, delta-sigma ADC and digital over-temperature detector with I^2C interface. It has a temperature accuracy of $-55°C$ to $+125°C$.

- *Pressure sensor:* BMP085 pressure sensor with pressure range 300 to 1100 hPa, low power, low noise device.

- *ADC:* AD7992 is a 12-bit ADC with fast conversion time (2 μs typically), 2 analog input channels, I^2C interface.

- *EEPROM:* PCA94S08 is an 8-kbit EEPROM with I^2C interface.

- *Real time clock:* PCA8565 is a low power CMOS real time clock and calendar with I^2C interface.

To compare between SPI and I^2C, SPI supports full duplex communication with higher throughput than I^2C. SPI communication is not limited to 8-bit words. As discussed earlier, it can utilize higher word sizes. Hence, we can send any message sizes with arbitrary content and purpose. The SPI interface does not require pull-up resistors. This results in lower power consumption. However, I^2C is simpler by having fewer lines which means fewer pins are required to interface to an IC. When communicating with more than one slave, I^2C has the advantage of in-band addressing as opposed to have a chip select line for each slave. I^2C also supports slave acknowledgement. This ensures the existence of the receiving device to the transmitter. This is not possible with SPI. In general SPI is better suited for applications that deal with longer data streams and not just words like address locations. Mostly longer data streams exist in applications working with a digital signal processor or analog-to digital converter. For example, SPI would be perfect for playing back some audio stored in an EEPROM and played through a digital to analog converter DAC. Moreover, since SPI can support significantly higher data rates comparing to I^2C, mostly due to its duplex capability, it is far more suited for higher speed applications reaching tens of megahertz. Since there is no device addressing involved in SPI, the protocol is a lot harder to use in multiple slave systems. This means when dealing with more than one node, generally I^2C is the way to go. The choice of SPI or I^2C depends on the following criteria.

- 2 or 3 wires are available
  - 2 I^2C
  - 3/4 SPI

- Speed requirement
  - 100–400 Kbits I^2C
  - 1 to 4 Mbit SPI

5.3 RS-232C

It is one of the oldest serial communication interface standards existing from 1960s. It can interface serial devices over a cable of length up to 25 metres and at data rates up to 38.4 Kbps. Though many other interfaces have been invented for serial transmission, because of its simplicity, RS-232C can still be used for designing embedded systems.

It is an unbalanced interface, i.e., the transmitted data bits have voltage levels with reference to local ground only. A logic high is –5 V to –15 V (typically, –12 V), a logic low is between +5 V and +15 V (typically +12 V). In the original terminologies of RS-232C, a logic high is called a *space*, whereas a logic low is called a *mark*. RS-232C was originally intended for connecting *Data Terminal Equipment* (DTE) with *Data Communication Equipment* (DCE). For example, a computer may be the DTE and the modem is a DCE. However, with different types of communication requirements, several interface requirements have also come up. For example,

- DTE-DTE: PC to a terminal
- DCE-DTE: Connect a workstation (with DCE interface) to a terminal
- DCE-DCE: Workstation to modem

Each of these requires a different type of cable. Thus, there exists different types of cables for RS-232C based interface. This is the major difficulty faced in interfacing using RS-232C. Table 5.1 shows the standard connections for both 25-pin and 9-pin connectors.

The handshaking between the transmitter and the receiver in RS-232C can be done either in hardware or in software. In hardware handshaking, when the transmitter wishes to send, it asserts RTS. It informs the receiver that there are some pending data with the transmitter. The receiver in turn asserts CTS when it is ready, indicating that the transmitter may send the data now. Software handshaking, also known as *XON/XOFF*, can be used when hardware handshaking is not possible. It uses two special characters *Ctrl-S* and *Ctrl-Q* to represent *request to suspend* and *clear to resume* respectively. Naturally, the transmitted data cannot

Table 5.1 Pin connections for RS-232C

Signal	Function	25-pin	9-pin	Direction
Tx	Transmitted data	2	3	From DTE to DCE
Rx	Received data	3	2	To DTE from DCE
RTS	Request to send	4	7	From DTE to DCE
CTS	Clear to send	5	8	To DTE from DCE
DTR	Data terminal ready	20	4	From DTE to DCE
DSR	Data set ready	6	6	To DTE from DCE
DCD	Data carrier detect	8	1	To DTE from DCE
RI	Ring indicator	22	9	To DTE from DCE
FG	Frame ground (chassis)	1	–	Common
SG	Signal ground	7	5	Common

have these characters embedded in it. Proper care should be taken to pre-process the data to handle these situations.

Many a time it is needed to power the interfaced device from the host directly. RS-232C can be utilized for this purpose. For example, the lines like RTS and DTR may not be used in many RS-232C applications. In such cases, either of these lines can be utilized as the power input to a voltage regulator and thereby provide the interfaced device with power. To switch ON or OFF the device, the host can raise or lower the RTS signal respectively through software.

5.4   RS-422

RS-422 is a balanced transmission system, that is, signal values are not referenced to local ground. It uses a *twisted pair* or *differential pair* to represent the logic level. Any noise introduced will affect both the wires in the pair, and thus, the difference between them will be less affected. Due to this *common mode rejection*, RS-422 can carry data over a longer distance (up to about 1200 metres) at higher data rates with greater noise immunity than RS-232C. The voltage difference between the pair of wires varies between ± 4 V and ± 12 V, thus being compatible to RS-232C.

5.5   RS-485

It is a low-cost networking version of RS-422. It is being used commonly in many industrial applications like data acquisition and control. It is capable of internetworking multiple transmitters and receivers on the same network. Using the default RS-485 receivers with an input resistance of 12 kΩ, it is possible to connect 32 devices to the network. The high-resistance RS-485 inputs allow this number to be expanded to 256. Repeaters are also available to increase the number of nodes to several thousands, spanning multiple kilometres. The interface also does not require intelligent network hardware: the implementation on the software side is not much more difficult than with RS-232. This is why it is popular with computers, PLCs, microcontrollers and intelligent sensors in scientific and technical applications.

Figure 5.7 shows the general topology of RS-485. N nodes are connected in a multipoint RS-485 network. For higher speed and longer lines, the termination resistances are necessary at both ends to eliminate reflections. The network should be designed as one line with multiple drops.

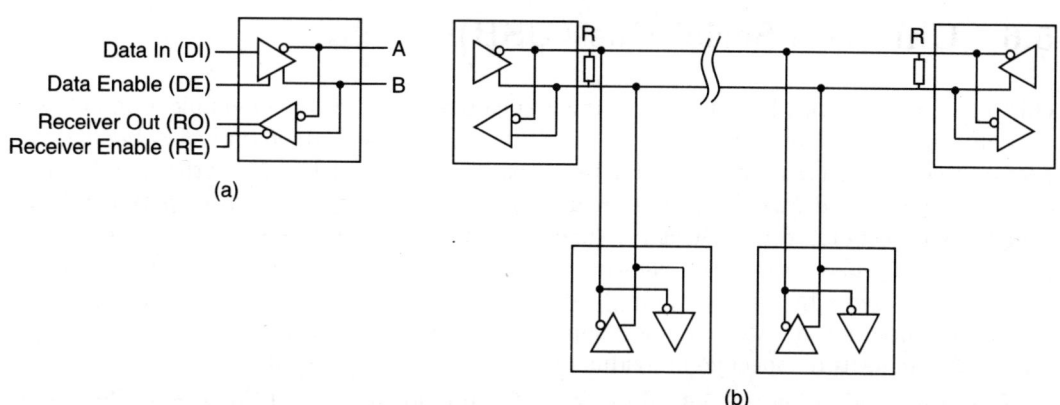

Fig. 5.7 (a) RS-485 transceiver, (b) Half-duplex RS-485 network topology.

As far as the functionality is concerned, by default, all the senders on the RS-485 bus are in tri-state with high impedence. In most higher level protocols, one of the nodes is defined as master which sends queries and commands over the bus. All other nodes receive data. Depending upon the information in the sent data, zero or more nodes on the line respond to the master. Bandwidth utilization is almost 100%. However, there is RS-485 implementation in which every node can start a data session on its own. Due to the chance of data collision, only 37% of the bandwidth will be effectively used. It is necessary that there is error detection implemented in the higher level protocol to detect data corruption and resend the information. There is no need for the senders to explicitly turn on/off the RS-485 driver. The drivers automatically return to their high-impedance tri-state within a few microseconds after the data has been sent. Thus, it is not needed to have delays between the data packets. Table 5.2 shows a comparison between RS-232 and RS-485.

Table 5.2 Comparison between RS-232 and RS-485

Specification	RS-232	RS-485
Mode of operation	Single ended	Differential
Number of drivers and receivers	1 driver, 1 receiver	32 drivers, 32 receivers
Maximum cable length	25 m	1200 m
Data rate	20 Kb/s	10 Mb/s
Driver output voltage	±25 V	−7 V to +12 V
Signal level (Loaded Min.)	±5 V to ±15 V	±1.5 V
Signal level (Loaded Max.)	±25 V	±6 V
Receiver input voltage range	±15 V	−7 V to +1 V
Receiver input resistance	3 K to 7 K	More than 12 K

5.6   Universal Serial Bus (USB)

One basic problem with RS-232C and similar such old standards of communication is that of difficulty in making two devices talk to each other. They need right type of cable and connectors, manual intervention to coordinate the parameters like data rate, parity, handshaking, etc. *Universal Serial Bus* (USB) has emerged as a solution to interconnect peripherals and computers in a standard way. It uses a single, standardized interface socket and provides improved plug-and-play capabilities allowing devices to be connected and disconnected without rebooting the computer (hot swapping). It also provides power to low consumption devices without the need for an external power supply and allowing many devices to be used without requiring manufacturer specific, individual device drivers to be installed.

USB 1.0 specification was introduced in 1995, promoted by Intel, Microsoft, Philips etc. The revised version *USB 1.1* came out in 1998. USB 1.1 supports data rates of 12 Mbps (*Full Speed*) and 1.5 Mbps (*Low Speed*). *Full Speed* was intended for high speed devices, such as disk drives, whereas, *Low Speed* for slower devices, such as joysticks. USB 2.0 was released in 2000 and ratified in 2001. It provides higher data transfer rate of 480 Mbps, while being fully compatible with USB 1.1. USB 3.0 has been published in 2008 and devices having this interface started reaching the market by 2010. Apart from being backward compatible with USB 2.0, USB 3.0 includes a new higher speed bus, called *SuperSpeed* in parallel with USB 2.0 bus. It has data transfer rate of upto 5 Gbps. Two-way communication is possible in USB 3.0. Using the *SuperSpeed* transfer, full duplex communication can be achieved, while the earlier USB 1.1 and USB 2.0 are only half duplex.

USB has an asymmetric design, consisting of a host, a multitude of downstream USB ports, and multiple peripheral devices connected in a tiered-star topology. Additional USB hubs may be included in the tiers, allowing branching into a tree structure, subject to a limit of five levels of tiers. One such structure has been shown in Fig. 5.8. A USB host may have multiple host controllers. Each host controller may provide one or more USB ports. Up to 127 devices

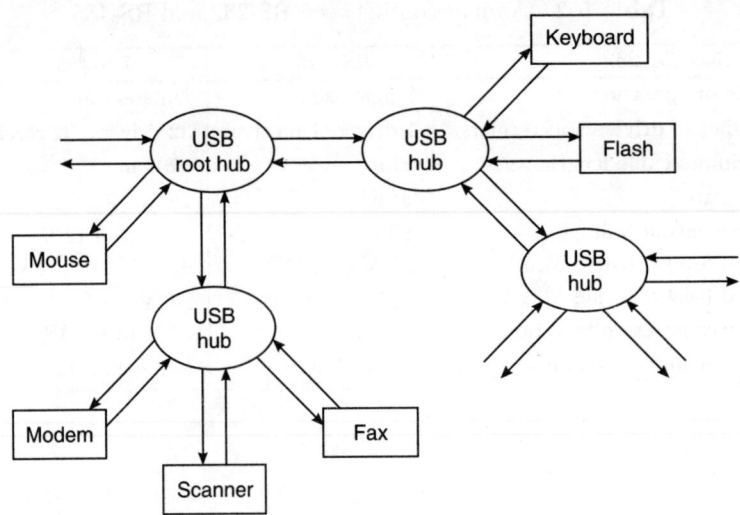

Fig. 5.8 Tiered USB tree structure.

including hub devices can be connected to a single host controller. Each host controller has a *root hub*. *Sharing hubs* are also possible, allowing multiple computers to access the same peripheral device(s). A single USB device may consist of several logical sub-devices, referred to as *device functions*. For example, a webcam (video device function) with a built-in microphone (audio device function).

The host regularly polls hubs for their status. When a new device is plugged into a hub, the hub advises the host of its change of state. The host in turn issues a command to enable and reset the port. The device responds and the host collects information about the device. Based on this retrieved information, the host operating system determines the software driver to be used for the device. A unique address is assigned to the device. On the other hand, when a device is unplugged, the hub advises the host about the change in state, when it is polled by the host. The host, in turn, removes the device from its list of available resources. This process of detection and identification of USB devices by a host is called *bus enumeration.*

USB device communication is based on *pipes* (logical channels). Pipes are connections from the host controller to a logical entity on the device, known as *endpoint*. A USB device can have up to 32 active pipes—16 into the host controller, 16 out of the host controller. Each pipe is unidirectional—an endpoint can transfer data in one direction only. Endpoints are grouped into *interfaces*, each interface is associated with a single device function. Endpoint zero is used for device configuration, and thus not associated with any interface. Figure 5.9 shows some USB endpoints.

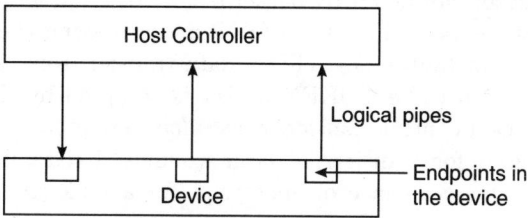

Fig. 5.9 USB endpoints.

USB supports various *device classes*. Devices attached to the bus can be full-custom devices, requiring a full-custom device driver, or may belong to a device class. The classes define the expected behaviour of the device and interface descriptors, so that the same device driver can be used for all such devices belonging to a class. An operating system is expected to implement all device classes to provide generic drivers for any USB device. Some of the common such device classes are noted in Table 5.3.

Four different types of data transfer can take place in USB. These are:

- Control transfer—to configure the bus and return status information.
- Bulk transfer—to move data asynchronously. It is bidirectional.
- Isochronous transfer—to move time critical data to an output device. This is unidirectional transfer without any error check.
- Interrupt transfer—to receive/transmit data at regular intervals, ranging from 1 to 255 milli seconds.

There are two types of pipes in USB communication—stream pipes and message pipes. A stream pipe is a unidirectional pipe connected to a unidirectional endpoint. Such a pipe

Table 5.3 USB device classes

Class	Purpose
Audio	Audio and music devices, sound systems
Chip/Smart card interface devices	Smart card devices
Common class	Generic devices
Communication devices	Modems, telephones, network interfaces
HID	Human Interface Devices such as mouse and keyboard
Hub	USB hub
IrDA	Infrared devices
Mass storage	Hard disks, CDROMs, DVD-ROMs
Monitor	Computer monitors and display devices
Physical interface devices	Joysticks and other similar devices
POS terminals	Point-of-sale devices such as cash registers
Power	Devices with power control or monitoring
Printer class	Printers
Imaging class	Scanners and cameras

is used in isochronous, interrupt, and bulk transfers. On the other hand, a message pipe is bidirectional, connected to a bidirectional endpoint and used exclusively for control transfers. To start a data transfer session, the host sends a TOKEN packet to the control endpoint of the device. The TOKEN packet can be an IN packet or an OUT packet. If the direction of data transfer is from the host to the endpoint, an OUT packet having the desired device address and endpoint number is sent by the host. For a data transfer from a device to the host, the host sends an IN packet. Once the TOKEN packet is accepted by the device, data transfer can start. This constitutes a control transfer, it uses message pipes.

A bulk transfer is used for transferring large volume of data in a unidirectional fashion using stream pipes. The sequence of operations for a bulk transfer has been shown in Table 5.4.

Table 5.4 Operation sequence for bulk transfer

Step No.	Device to Host	Host to Device
1	Host sends IN TOKEN packet to Device	Host sends OUT TOKEN packet to Device
2	Device sends bulk data to Host via data packet	Host sends bulk data to Device via data packet
3	Host sends acknowledgement to Device via a handshake packet	Device sends acknowledgement to Host via a handshake packet

The initial TOKEN packet provides the following information:

- Available bandwidth.
- Retry of transfer on failure.
- Guaranteed delivery of data but not bandwidth and latency.

No data content structure is specified. Low speed devices do not have bulk transfer endpoints.

The maximum data payload size for full speed is 8, 16, 32, or 64 bytes. For high speed (USB 2.0 onwards), it is 512 bytes.

Isochronous transfer provides guaranteed access of USB bandwidth with bounded latency. It uses stream pipe, and hence unidirectional. There is no retry in case of failure in transmission. Only full and high speed devices can do isochronous transfer. The transfer consists of TOKEN packet and data packet, but no handshake. Maximum packet size for full speed endpoints is upto 1023 data bytes, for high speed endpoints, it is 1024 data bytes. Many audio and video class devices use isochronous endpoints.

Interrupt transfer is designed for devices requiring to send or receive data infrequently but with bounded service period. Keyboard is a typical example of such a device. It provides guaranteed minimum service period and retrial on failure. Maximum data payload size is upto 8 bytes for low speed devices and upto 64 bytes for full speed devices, in case of interrupt transfer.

5.6.1 Physical Interface

USB uses a 4-wire shielded cable. Data are transmitted over a twisted pair data cable with 90 ohm $\pm 15\%$ impedance, in a differential manner. Transmitted signal levels are 0.0–0.3 V for low and 2.8–3.6 V for USB 1.1. For USB 2.0, the voltage levels are ± 400 mV. The USB wires are as noted in Table 5.5. The line *VCC* can provide power to low-powered devices. Such devices are known as *bus-powered devices*. For USB 2.0, the voltage level can be 5V$\pm 5\%$, for USB 3.0, the voltage can be in the range of 4.45 V to 5.25 V. A unit load is 100mA in USB 2.0 and 150 mA in USB 3.0. A device can draw a maximum of 5 unit loads (500 mA) from a port in USB 2.0 and 6 unit loads (900 mA) in USB 3.0. Some high-speed external disks require higher current. Such devices often use a Y-shaped cable having two USB connectors to be plugged into the host, thus drawing larger current.

Table 5.5 Wires in USB interface

Connector pin	Signal	Purpose	Wire colour
1	VCC	USB device power (+5V)	Red
2	D–	Differential data line	White
3	D+	Differential data line	Green
4	GND	Power and signal ground	Black

The maximum length of a standard USB cable is 5.0 metres, depicted by the maximum allowed round-trip delay of about 1500 ns. If a USB device does not respond within this maximum allowed time, the host assumes the command to be lost. However, using active extension cables (bus powered hubs), the cable length can be extended further. USB 3.0 does not directly specify any maximum cable length, requiring only that the cables meet an electrical specification.

5.6.2 USB Connectors

USB uses two different connection types. The connection from a device back to a host is called an *upstream connection*, while a connection from the host out to a device is called a *downstream connection.* Different connectors are used at the ports of these connections. The general features of these connectors are as follows:

1. The connectors are inexpensive to manufacture.

2. *Usability:*

   (a) It is difficult to connect USB connectors incorrectly. They cannot be plugged-in upside down.
   (b) By specification of USB, only a moderate force is needed for insertion/removal of the connectors.
   (c) It is not possible to create a cyclical network by using the connectors as they are incompatible.

3. *Safety:*

   (a) The connectors are robust. The electrial contacts are protected by an adjacent plastic tongue, and the entire connecting assembly is further protected by an enclosing metal sheath.
   (b) The enclosure also ensures that there is a moderate degree of protection from electromagnetic interference afforded to the USB signal.

4. USB specifies low tolerances for compliant USB connectors, intending to minimize incompatibilities.

Upstream connection uses *Series A* connectors, while downstream connection uses *Series B* connectors. The plug and receptacle for Series A and Series B differ from each other, making it impossible to fit a Series A plug into a Series B receptacle and vice versa. Figure 5.10 shows the connectors. Naturally, a Series A receptacle is found on a host or hub, and a Series A plug is at the end of the cable that attaches to the host or hub. On the other hand, a Series B receptacle is found on a USB device, whereas a Series B plug is at the end of the cable coming downstream from a host or hub. Figure 5.11 shows the situation. The USB hub will have both types of receptacles, as shown in Fig. 5.12.

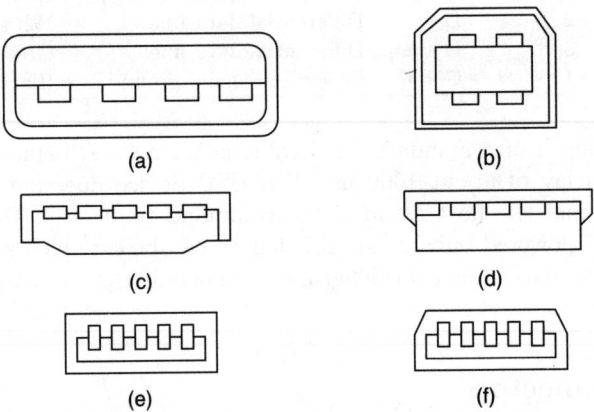

Fig. 5.10 USB 1.1 and 2.0 connectors: (a) Standard-A (b) Standard-B, (c) Mini-A, (d) Mini-B, (e) Micro-A, (f) Micro-B.

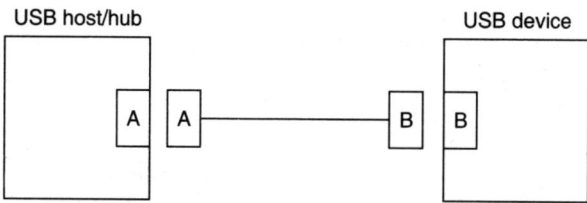

Fig. 5.11 USB connectors and cable.

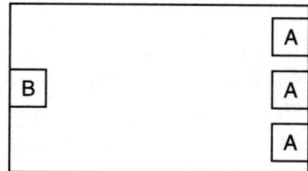

Fig. 5.12 Receptacles on a USB hub.

In Fig. 5.10 there are three different types of USB plugs—standard, mini, and micro. The standard connectors have 4 pins, while the mini and micro connectors have 5 pins. In such connectors, the fifth pin is the ground pin. The fourth pin stands for ID. It permits distinction of host connection from the slave connection. For the host, it is connected to the signal ground while for the slave, it is not connected.

USB 3.0 connectors

The standard type A USB 2.0 connector remains largely unchanged in 3.0. The 3.0 connector has five additional pins which are hidden internally. The extra pins allow for SuperSpeed transfer. However, the shape, size, and appearance of the USB 2.0 connector remains the same. Many manufacturers use blue-coloured plastic to distinguish a 3.0 connector from a 2.0 connector. The type B connector is changed a bit in 3.0. Compared to the 2.0 connector, the 3.0 type B connector has an extra protruding from the top of the connector. The added area has five additional pins for SuperSpeed data transfer. As a result, USB 2.0 connectors are compatible with USB 3.0 ports, but not vice versa. USB 3.0 standard A and B pin configurations are shown in Table 5.6.

Charging ports

Two new USB ports, namely *charging ports* have been introduced by the *USB Battery Charging Specification* of 2007. A standard downstream port can provide a maximum current of 100 mA after digital negotiation with the host. The charging ports can provide current above 0.5 A without negotiation. There are two types of charging ports—*Charging Downstream Ports* (CDP) and *Dedicated Charging Ports* (DCP). CDP supports data transfer as well, while DCP supports only charging. For a DCP port, $D+$ and $D-$ pins are shorted. For a CDP port, current may now exceed 900 mA during high speed data transfer, to avoid interference. A DCP port may have 0.5 to 1.5 A current rating. There is no upper limit on the current of CDP port, as long as the connector can handle it. Several hosts (such as, laptops) contain yellow coloured *Sleep-and-Charge* ports. They can charge the USB connected device even if

Table 5.6 USB 3.0 Standard A and Standard B connectors

Pin	Colour	Signal Name		Description
		A Connector	B Connector	
1	Red	VBUS		Power
2	White	D−		USB 2.0 differential data line
3	Green	D+		USB 2.0 differential data line
4	Black	GND		Power ground
5	Blue	StdA_SSRX−	STD_ASSTX−	SuperSpeed transmitter differential data line
6	Yellow	StdA_SSRX+	STD_ASSTX+	SuperSpeed transmitter differential data line
7	Shield	GND_DRAIN		Signal ground
8	Purple	StdA_SSTX−	STD_ASSRX−	SuperSpeed receiver differential data line
9	Orange	StdA_SSTX+	STD_ASSRX+	SuperSpeed receiver differential data line
Shell	Shell	Shield		Connector metal

the host/computer is turned off. The battery of the computer is used to provide the charging current through the yellow port.

5.7 Infrared Communication—IrDA

IrDA is the infrared transmission standard commonly used in computers and peripherals. It stands for *Infrared Data Association*, a consortium of more than 150 companies that maintain and develop the standard. It uses IR LEDs for communication at a wavelength of 870 ± 30 nm. With the increasing use of mobile equipments, importance of wireless digital links has increased manifold. Infrared radiation, as a medium for short-range communication, offers several significant advantages over RF transmission, especially for short-range, low-power, high data rate connection. High-speed infrared transceivers are available at low cost. It may be noted that *Bluetooth* links are up to five times more expensive to integrate than infrared links. The infrared spectral region offers virtually unlimited bandwidth that is unregulated worldwide. Like ordinary (visible) light, infrared can penetrate glass, but not walls or other opaque barriers. Thus infrared transmissions are confined to the originating room.

An *IrDA* infrared connection is established solely by a directed infrared beam. The IrDA transmitter beams out its transmission at an angle of 15 to 30 degree, either side of the line of sight (Fig. 5.13). The receiver has a viewing angle of 15 degree either side of its line of sight. The range of line of sight connection is at least 1 m, but typically 2 m can be reached. *Bit Error Ratio* (BER), defined as the number of incorrectly transmitted bits over the number of correctly transmitted bits, is 10^{-9} with the maximum level of surrounding illumination 10 klux (daylight). The initial IR communication takes place at 9600 bps, and devices negotiate the data rate up or down, depending upon the capabilities and need. However, unlike RS-232C type of connections, the user does not need to set, or know the bit rate. There is also a low power version available with a limited range of 20 cm, but at 10 times less power consumption. IrDA connections are suitable for point-to-point and even point-to-multipoint. It is very popular with portable devices, such as notebooks, handheld computers, digital cameras, etc.

At lower data rate, IrDA uses the bit encoding scheme *Return-to-Zero* (RZ) to transmit data. In this scheme, a frame consists of a transmission interval, that is divided into subintervals representing individual bits. Speeds for IrDA v1.0 range from 2400 to 115200 kbps.

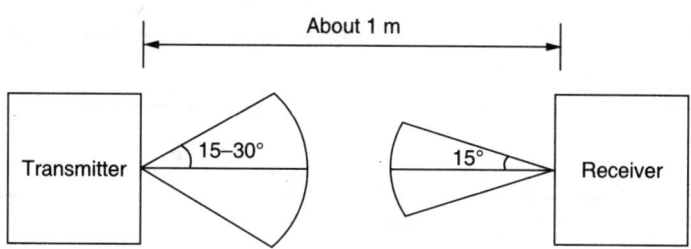

Fig. 5.13 The angles and the distance in IrDA transmission.

At this speed, logic 0 is represented by a pulse that is 3/16 the width of a bit subinterval, while a logic 1 is represented by the absence of a pulse (Fig. 5.14(a)). Data format is same as that for asynchronous serial transmission with a start bit at the beginning (shown in Fig. 5.15). In IrDA v1.1, for the transmission speeds of 0.576 and 1.152 Mbps, one quarter pulse duration is used, instead of 3/16. A starting sequence of 01111110 marks the beginning of a packet (frame). The packet format is shown in Fig. 5.16. It consists of two start words followed by address, data, cyclic redundancy code (CRC) bits, and an end flag. The data part of the packet uses bit stuffing (transmitter inserts a 0 after every five 1's in data bits to avoid false end-of-packet, receiver destuffs the inserted bit). The 16-bit CRC polynomial is $x^{16} + x^{12} + x^5 + 1$. Minimum of three STO fields are necessary between back-to-back frames. Least significant bit is transmitted first. Eight bits are used to encode each character. An abort of transmission is indicated by a minimum of seven consecutive 1's.

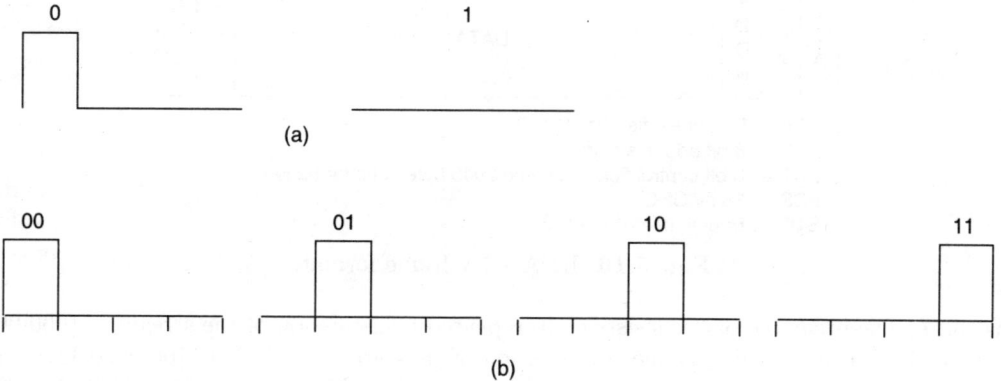

Fig. 5.14 (a) RZ coding (b) 4PPM coding.

At the higher data rate of 4 Mbps, *Pulse Position Modulation* (denoted as *4PPM*) is used to represent the bits. In this scheme, a data symbol duration (*Dt*) is defined which is further divided into four equal-length time slices called *chips* or *cells* (*Ct*), with $Dt = 500$ ns and $Ct = 125$ ns. Each data symbol represents two bits of payload data, as shown in Fig. 5.14(b). For example, bits 00 would be transmitted as a sequence 1000, bits 01 as 0100, bits 10 as 0010, and bits 11 as 0001. The major advantage of 4PPM is that only half of the LED flashes are needed compared to the previous modulation techniques. Thus, data can be transferred two times faster. A 4 Mbps data packet consists of the following fields: a 64-cell preamble, 8-cell start flag, frame body, a 4-byte CRC and an 8-cell stop flag. This has been shown in

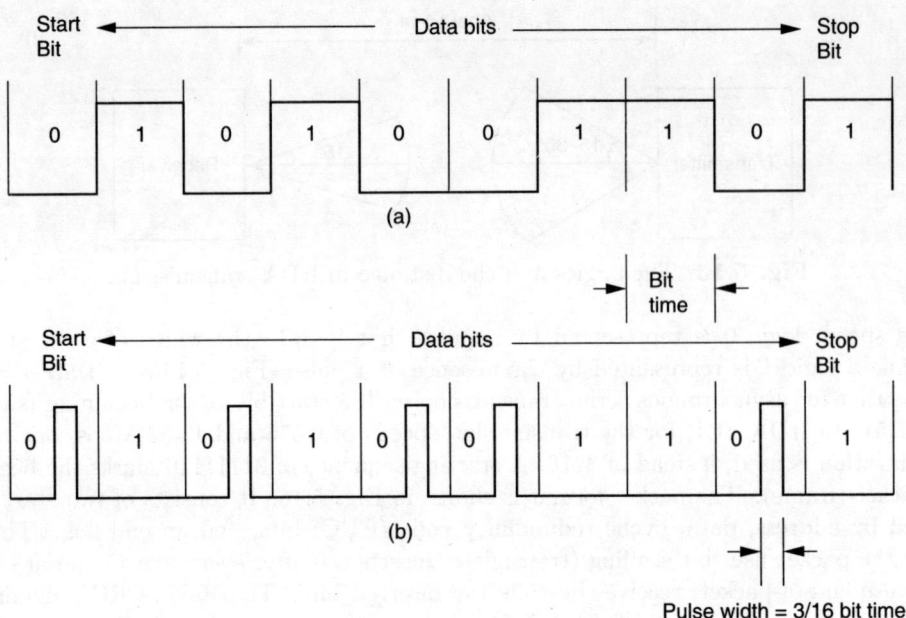

Fig. 5.15 (a) Asynchronous transmission frame. (b) IrDA frame.

S T A	S T A	A D D R	DATA	F C S	S T O

STA   : Beginning flag 01111110
ADDR : 8 bit address field
DATA : 8 bit control field plus upto 2045 bytes of information
FCS   : 16 bit CRC
STO   : Ending flag 01111110

Fig. 5.16 IrDA 1.1 v frame format.

Fig. 5.17. The preamble part consists of 16 repeated transmission of the stream $<$ 1000 0000 1010 1000 $>$. The start flag is one transmission of the sequence $<$ 0000 1100 0000 1100 0110 0000 0110 0000 $>$. The stop flag is one transmission of $<$ 0000 1100 0000 1100 0000 0110 0000 0110 $>$. The CRC32 polynomial is given by, $x^{32} + x^{26} + x^{23} + x^{22} + x^{16} + x^{12} + x^{11} + x^{10} + x^8 + x^7 + x^5 + x^4 + x^2 + x + 1$.

Preamble 64 cells	Start 8 cells	Frame body Upto 2050 bytes	CRC 4 bytes	Stop 8 cells

Fig. 5.17 A 4 Mbps data packet of IrDA.

It may be noted that the pulse modulation aids in error-free transmission in IrDA. To distinguish a signal from the surrounding illumination and noise, the receiver needs the signal with highest possible power. However, IR LEDs cannot transmit in full power for 100% of the bit time. Hence, usage of a pulse of narrower width ensures the transmitted power to be about 4 to 5 times higher than that in continuous illumination. Figure 5.18 shows the IrDA communication protocol layers. The layers within the stack can be divided into two groups—required protocols and optional protocols. The required protocol layers are the following:

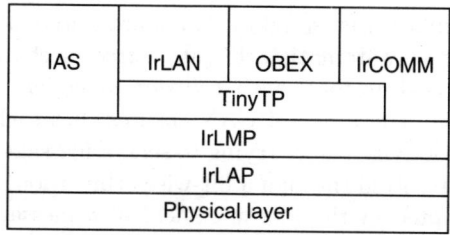

Fig. 5.18 Protocol layers of IrDA.

- *Physical layer:* It specifies the optical characteristics, encoding of data and framing for various speeds. There are two specific protocols: IrPHY1.0 for 2400-115200 bits/s, and IrPHY1.1 for 1.152-4.0 Mbps.
- *IrLAP* (Link Access Protocol): It establishes the basic reliable connection.
- *IrLMP* (Link Management Protocol): It multiplexes services and applications on the IrLAP connection.
- *IAS* (Information Access Service): It provides the list of services available on a device.

The use of optional protocols depends upon the particular application. The optional protocols are as follows:

- *TinyTP* (Tiny Transport Protocol): It provides per-channel flow control to keep things moving smoothly.
- *IrOBEX* (Object Exchange Protocol): It provides easy transfer of files and other data objects.
- *IrCOMM:* It provides serial and parallel port emulation. Thus, it enables existing applications using serial and parallel communications to use IR without change.
- *IrLAN* (Local Area Network): It allows local area network access via IR interface.

At lower data rates, protocol handling including packet formatting and error checking are usually done in software, however, for limited capacity of embedded processors, at higher speed, dedicated hardware components are used for IR interface.

5.8   Controller Area Network—CAN

Controller Area Network (CAN or CAN-bus) is a networking protocol originally designed for automotive electronics to allow microcontrollers and devices to communicate with each other. The complexity of automotive electronics has increased manifold with engine management systems, ABS braking, active suspension, lighting, air conditioning, security, etc. All these systems are also interconnected, performing a good amount of information exchange. A point-to-point connection between the systems will need a huge amount of wire—necessitating the usage of low-cost digital network. Another important feature of such a network is the highly noisy environment. With all the mechanical and electrical components on board, the 12 V supply to automotive electronics can have ±400 V transients. Thus the network must provide high noise immunity, as well as error detection and handling, with retransmission of failed packets.

CAN uses a broadcast, differential serial bus standard for connecting nodes (device units). The data transmission uses an automatic arbitration free mechanism. The protocol used is a *carrier-sense, multiple-access* protocol with *collision detection* and *arbitration on message priority* (CSMA/CD+AMP). The usage of CAN ensures that each node on a bus must wait for a prescribed period of inactivity before trying to send a message. *CD+AMP* implies that in case a collision occurs, it is resolved through a bit-wise arbitration, based on a preprogrammed priority of each message, noted in the identifier field of a message. When multiple devices transmit simultaneously, the one transmitting more *dominant bits* wins. Thus, it has higher priority. The node transmitting lower priority message will sense this and back off and wait. This is achieved by the assumption of '0'-bit to be *dominant* and a '1'-bit as *recessive*. The bus is physically an open-collector wired-AND connection. If a node transmits a dominant bit and another transmits a recessive bit, then the dominant bit wins. During arbitration, each transmitting node continually reads the bus and compare with the transmitted bit. In case a dominant bit is received, while the device had transmitted a recessive one, the device has lost the arbitration and stops transmitting the remaining bits. For example, consider two CAN messages with 11 bit ID (explained later) values 00000111111 and 00001111100, respectively. Till the transmission of fourth bit, both the devices continue. However, on the fifth bit the first message wins the arbitration as it transmits a dominating bit 0 while the second message contains 1—a recessive bit. The first message continues, while the second message is aborted, i.e., the node transmitting the recessive bit senses the situation and backs off. Thus, the higher priority message is never delayed. The lower priority node attempts retransmission 6 bit-clocks after the end of dominant message. Hence, allocation of message IDs plays a very important role in the system design. Each ID must be unique. The system designer must do this very carefully. Otherwise, if an urgent message with short deadline has a numerically high ID (lower priority), its transmission may be delayed by other messages with lower numerical IDs (higher priority), even if these messages may have longer deadlines.

A CAN network can be configured to work with two different message (frame) formats—the standard or base format (CAN 2.0 A) and the extended frame format (CAN 2.0 B). The base format supports 11 bits for the identifier field while the extended format supports 29-bit identifier. There are four different types of frames in CAN—*Data frame, Remote frame, Error frame,* and *Overload frame.*

Data frame

The data frame is used for data transmission. Table 5.7 shows the fields of a base data frame with 11-bit identifier while Table 5.8 shows the fields of an extended data frame.

Remote frame

In general, data transmission is performed on an autonomous basis from the source via data frame. However, in some cases, a destination node may request for data from a source by

Table 5.7 Base data frame structure

Field name	Length (in bits)	Purpose
Start-of-frame	1	Marks start of transmission
Identifier	11	Unique identifier for the message
Remote transmission request (RTR)	1	Dominant (0) (discussed later)
Reserved bit (r0)	1	Reserved bit (set to 0)
Data length code (DLC)	4	Number of bytes of data (0-8)
Data field	0–64	Data to be transmitted
CRC	15	Cyclic redundancy code
CRC delimiter	1	Must be recessive (1)
ACK slot	1	Transmitter sends 1 and any receiver can assert 0
ACK delimiter	1	Must be recessive (1)
End-of-frame	7	Must be recessive (1)

Table 5.8 Extended data frame structure

Field name	Length (in bits)	Purpose
Start-of-frame	1	Marks start of transmission
Identifier A	11	First part of unique identifier
Substitute remote request (SRR)	1	Must be recessive (1)—a simple place holder
Identifier extension bit (IDE)	1	Must be recessive (1)
Identifier B	18	Second part of unique identifier
Remote transmission request (RTR)	1	Dominant (0)
Reserved bits (r0, r1)	2	Must be dominant (0)
Data length code (DLC)	4	Number of bytes of data (0–8)
Data field	0–64	Data to be transmitted
CRC	15	Cyclic redundancy code
CRC delimiter	1	Must be recessive (1)
ACK slot	1	Transmitter sends 1 and any receiver can assert 0
ACK delimiter	1	Must be recessive (1)
End-of-frame	7	Must be recessive (1)

sending a Remote frame. The RTR bit is transmitted as dominant in data frame and recessive in remote frame. Also, the remote frame does not have any data field in it. The setting of RTR in this fashion ensures that in the very unlikely case of a data frame and a remote frame with the same identifier being transmitted at the same time, the data frame wins and the node that transmitted the remote frame gets the data immediately.

An error frame passes the error information in the network. It is transmitted by a node detecting an error condition in a message and causes all other nodes to transmit error frames as well. As a consequence, the source node automatically retransmits the message. Individual CAN controllers contains *error counters* to avoid repeated transmission of error frames by any node. Overload frames are similar to the error frames, as far as the format is concerned. It is transmitted by a node when it becomes too busy. It is used to provide extra delay between messages.

The CAN bus

Each node in a CAN consists of the following:

- A *host processor:* It determines the type of the messages received, their meaning, and also the messages the node wants to transmit.

- A *CAN controller:* It is a hardware unit working on a synchronous clock signal. It performs the operation of receiving and sending messages. On the receiver side, it stores the received bits from the bus until the entire message is available. The CAN controller then sends an interrupt to the host processor to inform it to collect the message. For sending messages, host controller stores the message to be transmitted into the CAN controller which then transmits bits serially through the bus.

- A *transceiver:* The transceiver may be integrated with the CAN controller. On the receiving side, it adapts signal levels from the bus to the levels that the CAN controller expects. It also possesses protective circuitry that protects the CAN controller. On the sending side, it converts the transmit bit signal received from CAN controller into a signal that can be sent over the CAN bus.

The structure of a CAN-bus has been shown in Fig 5.19. The High-Speed ISO 11898 Standard specification suggests both ends of the bus to be terminated by a 120 Ω resistor to avoid reflection. CAN-bus can transmit data at 1 Mbps at network lengths below 40 m with a maximum of 30 nodes. At lower data rates (for example, 125 Kbits/s), distances can be increased. For exmaple, at 125 Kbits/s, 500 m distance can be covered. The two signal lines in the CAN bus, CANH and CANL are passively biased at a quiscent voltage when no signal is being transmitted. In the dominant state, CANH assumes a positive voltage level compared to the quiscent level, while CANL assumes a negative value, creating a differential signal over the two lines.

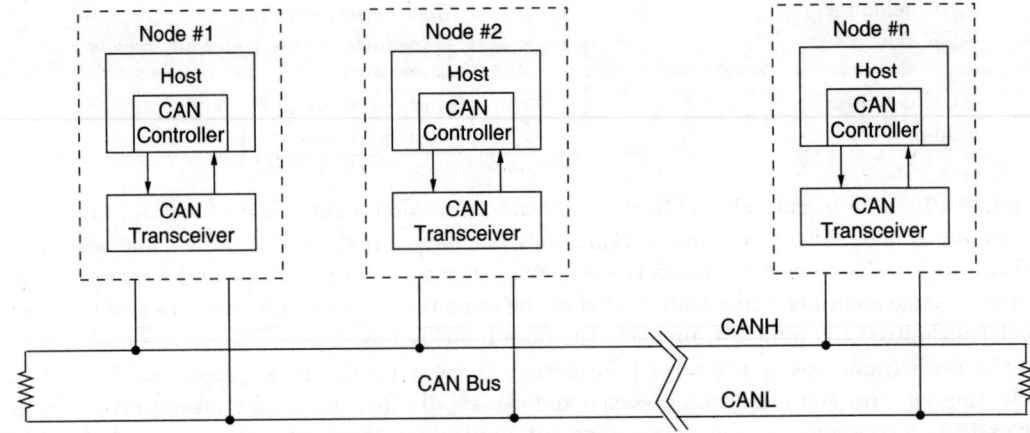

Fig. 5.19 CAN bus structure.

The CAN protocol consists of the following layers:

- *Physical layer:* It defines how the signals are actually transmitted. It includes signal level and bit representations with conversion, and maintenance of transmission medium.
- *Transfer layer:* It is the kernel of the CAN protocol responsible for bit timing and synchronization, message framing, arbitration, acknowledgement, error detection and signalling, fault confinement, etc.
- *Object layer:* It is responsible for message filtering, message and status handling.
- *Application layer:* This is responsible for differnt applications built on the underlying CAN protocol layers.

5.9 Bluetooth

Conceived initially by *Ericsson*, before being adopted by myriad of other companies, Bluetooth is a standard for a small, cheap radio chip to be plugged into electronic devices like computers, printers, mobile phones, etc. to enable wireless communication between them. It is named after the late tenth century king Herald Bluetooth, King of Denmark and Nórway, known for his unification of previously warring tribes from Denmark and Norway. Bluetooth likewise is intended to unify different technologies. It is intended to be an inexpensive, flexible, and robust replacement for short-range cables. It enjoys several advantages over infrared communication. First, infrared is a *line of sight* technology. Secondly, it is almost always a *one-to-one* technology.

Bluetooth networking transmits data through low-power radio waves at a frequency of 2.45 GHz (between 2.40 GHz–2.4835 GHz). The frequency band has been internationally ear-marked for use of industrial, scientific, and medical devices. Bluetooth devices avoid interfering with other systems by sending weak signals of about 1 mW making the range of Bluetooth transmission restricted to 10 metres. Supported data rates are 108/108 Kbps symmetric channel and 723/57 Kbps asymmetric channel. It also does not require a line-of-sight communication. Moreover, the walls cannot stop a Bluetooth signal, making the standard useful for controlling several devices in different rooms. Bluetooth can connect upto eight devices simultaneously. Still the devices do not interfere with each other. This is because of the *spread spectrum frequency hopping* used in Bluetooth. In this scheme, a device will use 79 individual, randomly chosen frequencies within a designated range, changing from one frequency to another on a regular basis. The transmitters change frequencies 1600 times every second. Similar to Wi-Fi (IEEE 802.11 standard), Bluetooth is also used for setting up networks, printing, or transferring files. However, Wi-Fi is used to establish a wireless local area network (WLAN), whereas, Bluetooth is intended for portable equipments and its applications (commonly known as wireless personal area network, WPAN). Wi-Fi uses the same radio frequencies as Bluetooth, but with higher power, resulting in higher bit rates and better range.

When Bluetooth-capable devices come within the range of one another, an electronic conversation takes place to determine whether the devices have data to share or whether one needs to control the other. No intervention from the user is necessary. Once the conversation has occurred, the devices form a *Personal Areā Network* (PAN) called a *piconet*. It has a major device and upto seven slave devices. A piconet is an ad-hoc computer network. Though only seven slave devices may be active, upto 255 further devices can be inactive, or parked, which a master device can bring into active status at any time. Two or more piconets may be connected together to form a *scatternet*, with some devices acting as a bridge by simultaneously playing the master role and slave role in one piconet.

Any Bluetooth device will transmit the following information on demand:

- Device name
- Device class
- List of services
- Technical information, such as device features, manufacturer, clock offset, etc.

Every device has a unique 48-bit address. However, instead of using this address, device names are used which can be conveniently set by the user. A typical Bluetooth protocol stack has been shown in Fig. 5.20. A brief description of the protocol follows:

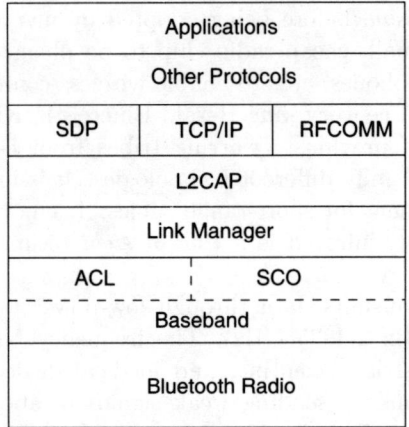

Fig. 5.20 Bluetooth protocol stack layers.

1. *Bluetooth radio:* It is a transceiver to send and receive signals from peer Bluetooth devices. The frequency band, modulation techniques, etc. have already been discussed earlier.

2. *Baseband:* It manages the physical channels and links, provides error correction, security, etc. It uses a *time division duplex* (TDD) scheme, alternating between transmit and receive. It handles two types of links—*Synchronous Connection-Oriented* (SCO) and *Asynchronous Connection-Less* (ACL). SCO is a circuit-switched, symmetric, point-to-point link between a master and a single slave in a piconet. SCO packets are never retransmitted. The link is mainly used for speech transmission. On the other hand, ACL is a packet-switched, point-to-multipoint link. While there can be upto three simultaneous SCO links, only a single ACL link can exist at a time. Packet retransmission is applied in ACL. A standard Bluetooth packet has a 72-bit *access code*, 54-bit *header* and upto 2745-bit *payload*. The access code is used for timing synchronization, offset compensation, paging and inquiry. Three different types of access codes are present—*Channel Access Code* (CAC) identifies a piconet, the *Device Access Code* (DAC) is used to page for devices and their responses and the *Inquiry Access Code* (IAC) is used for inquiry purposes. The header contains information about packet acknowledgement, packet

numbering (for out-of-order packets), flow control, slave address, error check for header. Payload can contain either voice and/or data.

3. *Link Manager:* It handles link setup, security and control. It also manages devices in different modes. A Bluetooth device can be in one of the four states—*active, sniff, hold, park.* In active mode, the device actively participates in listening the channel, receiving and transmitting packets. In sniff mode the slave device, instead of listening for every slot for master's message for it, sniffs on specified time slots only. Slave can sleep in other time slots, saving power. In hold mode, a device can temporarily not support ACL packets and go to low power sleep mode to make the channel available for other works. In park mode, the device does not participate in the piconet channel, but still remains synchronized to the channel.

4. *L2CAP: Logical Link Control and Adaptation Layer Protocol* resides in the data link layer and provides data services to the upper layers. L2CAP data packets sent/received to/from upper layers can be up to 64 Kbit long.

5. *SDP: Service Discovery Protocol* is used to discover which services are available and their characteristics.

6. *RFCOMM: Radio Frequency Communications* is a simple transport protocol that serves the purpose of cable replacement. It provides binary data transport and emulates EIA-232 (previously RS-232) control signals over L2CAP. Many Bluetooth applications use RFCOMM due its wide support in various operating systems. The applications using serial port to communicate can be easily ported to use RFCOMM.

Pairing/Bonding

When two Bluetooth devices come within a range of communication, they try to establish a connection between them, often completely unnoticed by the users. This process is known as *pairing* or *bonding*. The process may also be initiated by the user who explicitly requests the device to *add a Bluetooth device*. Version 2.1 onwards, pairing takes place via *Secure Simple Pairing* (SSP). It uses a public key cryptography framework to establish the bond. It has several modes of operation:

1. *Just works:* No user interaction is required. The device may ask the user to confirm the pairing process. It does not provide any protection against *man in the middle* (MITM).

2. *Numeric comparison:* It requires both the devices to have a display and at least one device capable of accepting a binary Yes/No user input. It displays a 6-digit numeric code on each display. The user should compare the numbers to ensure that they are identical and then confirm the pairing. It provides protection against MITM.

3. *Passkey entry:* The method requires at least one device to have a display and at least one device to have a numeric keypad. A 6-digit numeric code is displayed and the user enters the code through the keypad. If both devices have keypads, user of each device enters the code.

4. *Out of band:* It uses some external means of communication to exchange some information used in the pairing process. Pairing is performed using Bluetooth radio, as it requires this external information.

5.10   Conclusion

In this chapter, we have seen various interfacing techniques commonly used in embedded systems to interface a variety of I/O devices. A wide range of interfaces have been covered. While the simplest interface, such as SPI is good enough for small embedded systems, advanced systems consisting of multiple modules may interact between themselves using very advanced techniques like Bluetooth. Since a good number of applications of embedded systems can be found in automotives, the CAN interfacing plays a vital role there. A detailed discussion on each of these techniques will take volumes. Each interfacing standard has its associated chipsets already designed for ready usage in embedded applications. Thus, in this chapter, we have restricted ourselves to an overview of many such interfaces, so that, given an application in hand and the communication requirements between its modules, we can determine which interface will be the most suitable.

Exercises

5.1 Why does an embedded processor need to support different types of interfaces, compared to a desktop?

5.2 Describe how serial peripheral interface can be used for data transfer.

5.3 Give examples of devices with SPI interface.

5.4 Enumerate the role of CPOL and CPHA bits of SPI control register.

5.5 Assume that transferring data to/from a register from/to SPI data register in CPU takes 2 ms and in the device 10 ms. Transmission over MOSI/MISO lines is at a rate of 64 Kbps. Compute the time required to transfer 1 Kbyte: (i) from CPU to device, and (ii) from device to CPU.

5.6 Discuss the data transfer technique using IIC interface.

5.7 Give examples of devices with IIC interface.

5.8 What are the major advantages and disadvantages of RS-232 series of protocols?

5.9 Compare between RS-232 and RS-485.

5.10 How does "hot swapping" take place in USBs?

5.11 Design a tiered USB tree structure for a system with 40 USB devices.

5.12 What are the different types of data transfers possible in USB? Explain them.

5.13 Describe the physical interfacing wires in USB.

5.14 Why is it not possible to connect USB connectors wrongly? What facility does it provide to the overall design?

5.15 Describe different types of USB connectors. Why should a hub contain both type A and type B connectors?

5.16 Enumerate the features of USB 3.0.

5.17 Explain charging ports in USB.

5.18 Why is IrDA called a 'line-of-sight' communication?

5.19 What are the various encoding schemes used in IrDA? For the following bit-stream, show the encoded symbols for both RZ and 4PPM coding: (i) 0101011, and (ii) 1100110101.

5.20 Describe the protocol layers of IrDA.

5.21 Describe the CAN protocol. Which feature of it makes it suitable for embedded applications, particularly in automotives.

5.22 Explain automated arbitration and collision detection in CAN with an example.

5.23 Explain frame structures in CAN.

5.24 Enumerate the sturcture of a CAN bus.

5.25 What are the advantages of Bluetooth over IrDA?

5.26 How does the communication take place between devices in Bluetooth?

5.27 Explain Bluetooth protocol layers.

5.28 In the context of Bluetooth communication, explain pairing/bonding.

CHAPTER 6

Real-time Operating System

In the last few chapters, we have seen mainly the hardware platforms available for embedded system design. For small and simple embedded systems, it may be the case that the software portion is minimal. The hardware may not need any special mechanism to control and coordinate its components to realize the application(s). However, for moderate to complex embedded systems, normally there exists a good number of coordinating software processes. Depending upon the properties of such processes (like criticality, periodicity, deadline, etc.), a proper scheduling is necessary. This ultimately leads to the requirement of software to control the overall operation of the system. Such a software is the *operating system* (OS). It is the piece of software responsible for managing the underlying hardware and providing an efficient platform on which applications can be developed. While designing an embedded application to run on some host processor, it is thus necessary for the application developer to understand the policies followed by the underlying operating system. On the other hand, since the embedded systems are mostly real-time functions having continuous interaction with the environment, it is very much important that the OS also supports real-time tasks. In this chapter, we will see the features of real-time tasks and the policies followed by a real-time system to handle those tasks. This essentially involves the scheduling of such tasks.

6.1 Types of Real-time Tasks

Based on their time criticalities, real-time tasks can be classified into three categories. These are,

- Hard real-time tasks
- Firm real-time tasks
- Soft real-time tasks

6.1.1 Hard Real-time Tasks

Hard real-time tasks are those which must produce their result within a specified time limit or deadline. Missing the deadline implies that the task has failed. An example of such a system can be a plant consisting of several motors along with sensors and actuators. The plant controller senses various conditions through sensors and issues commands through the actuators. It is necessary that the controller responds to the input signals in a time-bound fashion. That is, on the occurrence of some event, the response from the controller should come within a pre-specified time limit. Examples of such signals may be inputs from fire-sensor, power-sensor, etc.

Another example can be a single (or a set of) robot(s) that performs a set of activities under the coordination of a central host. Depending upon the environment, a robot may need to take some instantaneous decisions, for example, on detecting an obstacle on its path of movement, the robot should find a way out to avoid a collision. A delay in the detection and reaction process may result in a collision, causing damage to the robot.

A very important feature of hard real-time system is its *criticality*. It means that a failure of the task along with a failure to meet the deadline will have a catastrophic effect on the system. Many of the hard real-time tasks are *safety-critical*. This is particularly true for the medical instruments monitoring and controlling the health of a patient. However, all hard real-time systems need not be safety-critical. For example, in a video game, missing a deadline is not that severe.

As far as scheduling of hard real-time tasks is concerned, we have to honour the deadline. Unlike in ordinary operating systems in which we try to complete tasks as early as possible to maintain a high throughput, there is no gain in finishing a hard real-time task early. As long as the task completes within the specified time limit, the system runs fine.

6.1.2   Firm Real-time Tasks

A firm real-time task also has an associated deadline, within which the task is expected to be completed. However, if it does not complete within this time (that is, deadline is missed), then also the system does not fail altogether. Only some of the results may need to be discarded. For example, in a video conferencing, video frames are sent over a network. Depending upon the properties of the network, some frames may arrive late, or may be lost. The effect is some degradation in the video quality for some time, however, that in most cases is tolerable.

The main feature here is that any result computed after the deadline is of no value, and thus discarded. As shown in Fig. 6.1, after the event has occurred, the utility of response is 100% if it occurs within the deadline. Beyond the deadline, utility becomes zero, and thus the result is simply discarded.

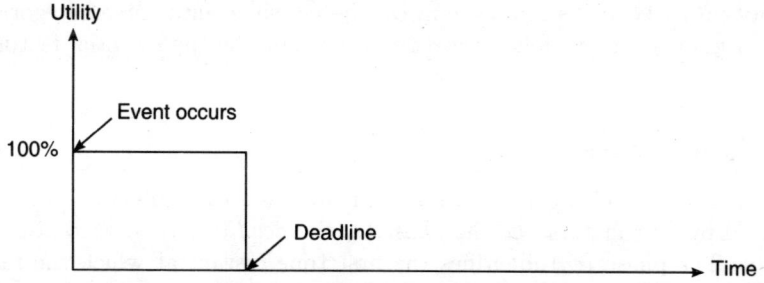

Fig. 6.1 Utility of result for firm real-time tasks.

6.1.3   Soft Real-time Tasks

The other category of real-time tasks are the soft real-time tasks. For such a task, there is a deadline, however, it is only expected that the task completes within the deadline. If the task does not complete within the deadline, still the system runs fine without any failure. Late

arrival of results does not force a total discarding of them. However, as the time passes, the
utility of the result drops, as shown in Fig. 6.2.

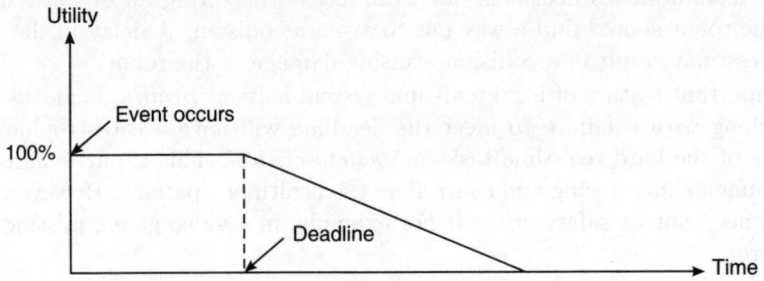

Fig. 6.2 Utility of result for soft real-time tasks.

A typical example of soft real-time task is the railway-reservation system, where it is only
expected that the average time needed to process the request for a ticket is small. If a particular
request for a ticket takes slightly larger time, still the system is acceptable, as nothing critical
happens to the system or the environment.

Another example of soft real-time system is web-browsing. After typing the URL, we expect
the page to arrive soon. However, we do not consider it as an exception if the process takes
slightly longer time.

6.2   Task Periodicity

An embedded system is generally designed to cater to a specific application, or a small number
of applications. Thus, it normally consists of a small set of tasks. For example, a monitoring
system continuously monitors its environment and takes appropriate actions based on inputs.
Such a system will generally consist of a set of tasks to be performed at some regular intervals
of time. However, there may be some other tasks which are not periodic in nature. Based on
this periodicity property, tasks in a system can be classified into three categories—*Periodic,
Sporadic* and *Aperiodic*. In the following section, we describe the essential features of each of
them.

6.2.1   Periodic Tasks

A *periodic task* repeats itself regularly after a certain fixed time interval. Such a task T_i can
be characterized by four factors—ϕ_i the *phase*, d_i the *deadline*, e_i the *execution time*, and p_i
the *periodicity*. The phase (ϕ_i) identifies the first time instant at which the task T_i occurs.
For a particular occurrence of the task, the deadline d_i is the time by which the task must
be completed, relative to the start time. The actual time required to execute the task in the
worst case is e_i, which must be smaller than d_i. Finally, p_i identifies the periodicity of the
task. Thus the task T_i can be represented by the four-tuple $< \phi_i, d_i, e_i, p_i >$. The situation
has been shown in Fig. 6.3.

It may be noted that most of the tasks in a real-time embedded system are periodic in
nature, parameters of these tasks can be computed beforehand. It is the responsibility of the
designer of the system to ensure the availability of proper resources at these times so that

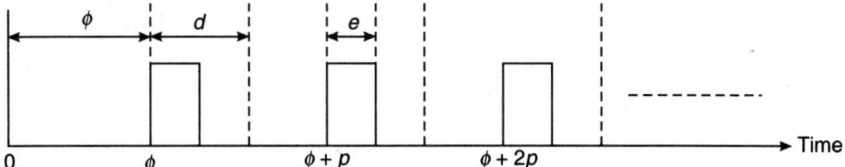

Fig. 6.3 Periodic task behaviour.

each instance of a periodic task can complete before its deadline. Sometimes, the deadline d_i is taken to be same as the periodicity p_i, as no further instances of the task can arrive in between.

6.2.2 Sporadic Tasks

As the name suggests, a *sporadic task* can arrive at any instant of time. However, after the occurrence of one instance, there is a minimum separation, only after which another instance of the task can arrive. Thus a sporadic task can be represented by the three-tuple $< e_i, g_i, d_i >$, where e_i is the worst case execution time of an instance of the task, g_i denotes the minimum separation between two consecutive instances of the task and d_i is the relative deadline. Examples of sporadic tasks are the interrupts which may be generated from different conditions, inside or outside the system. The criticality level may vary. A fire-alarm occurs sporadically, however, it is very critical and be handled within a fixed time interval. On the other hand, an I/O device interrupt may not be that critical. Figure 6.4 shows the behaviour of a sporadic task with execution time e, deadline d and minimum separation between two occurrences g. As shown in Fig. 5.4 the next instance of the task can arrive only after g time units of the occurrence of the previous instance.

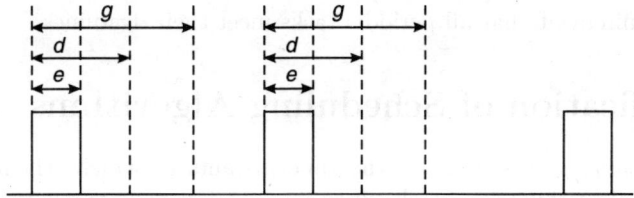

Fig. 6.4 Sporadic task behaviour.

6.2.3 Aperiodic Tasks

Aperiodic tasks are similar to the sporadic ones in the sense that both types of tasks can arrive at any random time. However, unlike sporadic tasks, there is no guarantee that another instance of the task will not arrive before a minimum amount of time expires. That is, successive instances of aperiodic tasks may occur even at consecutive time instants. Naturally, it is not possible to assign a tight deadline with these tasks. The deadline can be expressed either as an average value or some statistically expected value. In other sense, aperiodic tasks have to be soft real-time ones, as successive occurrence of instances of aperiodic tasks may lead to missing deadlines for some of them. These deadline misses can be tolerated by soft real-time tasks.

A typical example of aperiodic task is railway-reservation system. As booking is done over a large number of terminals, there is no periodic gap between two such tasks being initiated. At the same time, the deadline requirement is also an expected value only about the average time needed to process the ticketing requests.

6.3   Task Scheduling

The job of task scheduling refers to identifying the order in which tasks should execute in a system. Since most of the tasks in a real-time embedded system are periodic in nature, the real-time task scheduling algorithms mostly concentrate on periodic tasks. Sporadic and aperiodic tasks are handled as they occur, mostly on a case-by-case basis, without disturbing the deadlines of the already scheduled tasks.

A schedule for a given set of tasks is thus an assignment of time-frames and resources to individual tasks. A schedule is said to be *valid* if at every point of time, at the most, only a single task has been assigned to a processor, no task is scheduled before its arrival, the precedence constraints between the tasks are met, and the resource constraints are honoured. A *feasible* schedule ensures that all tasks meet their respective deadlines.

A large number of scheduling algorithms have been proposed in the literature. We will discuss about them shortly. The quality of a schedule is identified by a term called *processor utilization*. The processor utilization of a task is defined to be the fraction of time for which the processor is used by it. Thus, if the execution time of a task be e_i and the periodicity be p_i, then utilization, $u_i = e_i/p_i$. For a set of n tasks, the processor utilization is given by the sum of utilizations for individual tasks, that is, overall utilization, $U = \sum_{i=1}^{n} (e_i/p_i)$.

A good scheduling algorithm should lead to 100% processor utilization. However, as we shall see later, it is not always possible for a set of real-time tasks to ensure 100% utilization along with the requirement that all periodic tasks meet their deadlines.

6.4   Classification of Scheduling Algorithms

Based upon *scheduling points* (that is, the time instants at which scheduling decisions are taken), scheduling algorithms can be classified into the following categories:

- Clock driven scheduling
- Event driven scheduling
- Hybrid scheduling

In a *clock driven scheduling*, the scheduling points are the interrupts received from a periodic clock. There are two types of clock driven schedulers:

- Table driven
- Cyclic

On the other hand, *event driven schedulers* respond to different events in the system. They perform scheduling when an instance of a task has finished executing. There are mainly three types of event-driven schedulers:

- Simple priority scheduling
- Rate monotonic scheduling (RMS)
- Earliest deadline first (EDF) scheduling

In the following section, we give the details of each of these scheduling strategies.

6.5   Clock Driven Scheduling

As the name suggests, a clock driven scheduler works in synchronism with a clock/timer signal. The timer periodically generates interrupts. On receiving an interrupt, the scheduler is activated which then takes a decision about the process to be scheduled next. Since the set of tasks, their periodicity values, execution times required and deadlines are known beforehand, it is possible to precompute the processes to be scheduled at various clock interrupts. This helps in making the decisions simple, and thus the overall scheduler design is simplified. However, the major drawback of this type of schedulers is their inability to handle aperiodic and sporadic tasks, as their exact arrival times cannot be predicted, which implies that their scheduling points cannot be determined. This type of schedulers are therefore known as *static schedulers*.

6.5.1   Table Driven Scheduling

A table driven scheduler uses a precomputed table that stores the tasks to be run at different clock intervals. For example, consider a set of tasks $ST = \{T_1, T_2, T_3\}$ with the associated parameters as shown in Table 6.1. A possible schedule for this set of tasks can be as shown in Table 6.2. It is assumed that at the time instant 0, all three tasks have arrived. Next instance of T_2 will arrive at time instant 3. It is easy to see that the arrival pattern of task instances will repeat itself from the 12th instant of time, the least common multiple (LCM) of the periodicities of individual tasks. Thus, in the schedule table (Table 6.2), it is sufficient to store the number of entries equal to the LCM of the periods of the tasks. It can be verified that even if the tasks do not arrive exactly at time 0 to start with, that is, they possess a non-zero phasing, still the periodicity equal to the LCM holds. This LCM determining the size of the schedule table is called a *major cycle*. Instances of tasks occur in identical sequence in each major cycle.

Table 6.1   Example task set with parameters

Task id (T_i)	Execution time (e_i)	Periodicity (p_i)
T_1	2	6
T_2	1	3
T_3	4	12

6.5.2   Cyclic Scheduling

A major problem with table-driven scheduling is the size of the schedule table. If the LCM of periods is a large number, we need to have so many slots in the table, even if the execution times of individual tasks are high. In cyclic scheduling, the major cycle (equal to the LCM of periodicity of tasks) is divided into a number of equal sized *minor cycles* or *frames*. One or

Table 6.2 Schedule for Table 6.1

Time instant	Tasks arrived	Task scheduled
0	T_1, T_2, T_3	T_1
1		T_1
2		T_2
3	T_2	T_2
4		T_3
5		T_3
6	T_1, T_2	T_3
7		T_3
8		T_2
9	T_2	T_1
10		T_1
11		T_2
12	T_1, T_2, T_3	T_1
$\vdots$	$\vdots$	$\vdots$

more frames are allocated to individual tasks. The situation has been shown in Fig. 6.5, in which a major cycle has been divided into four equal-sized frames, F_1, F_2, F_3, F_4. Now, three tasks T_1, T_2, and T_3 have been allocated to various frames within the major cycle. The task T_1 has been assigned to two frames F_1 and F_3, whereas the tasks T_2 and T_3 have been assigned to frames F_2 and F_4, respectively. The scheduler gets up at individual frame boundaries via interrupts from a periodic timer and performs the switching of tasks that may be necessary between successive frames. The schedule table stores the tasks to be run for different frames, thus, the size of the table is reduced to be equal to the number of frames.

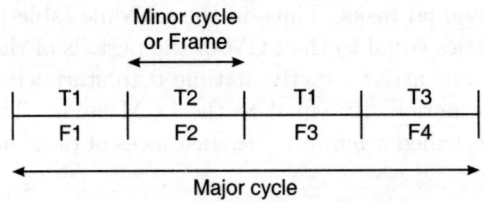

Fig. 6.5 Major cycle and frames.

The choice of frame size is a very important design parameter for cyclic schedulers. A large frame size implies lesser number of frames in a major cycle, however, it may lead to wastage of CPU time as for most of the tasks, execution time may be much smaller than the frame size. The constraints behind selection of a frame size are the following:

1. **Minimize context switching:** The frame size be such that even the largest task can complete within a frame. Otherwise, multiple frames need to be allocated to a task. Since the scheduler needs to be invoked at each frame boundary, it will necessitate a good number of context switches between the scheduler and the task. These times are wasted.

2. **Schedule table minimization:** Since the schedule table holds information for each of the frames, a large frame size is advisable, as it means lesser number of frames. Moreover,

the frame size be such that the major cycle is an integral multiple of it. Otherwise, storing information for one major cycle is not sufficient for scheduling.

3. **Satisfaction of task deadlines:** This is one of the very crucial issues in determining frame size. If proper care is not taken, it may so happen that a task arrives just after the start of a frame. Since scheduling is performed only at frame boundaries, it cannot be scheduled at least till the beginning of next frame. But, by that time, the task deadline may be very close, resulting in missing deadline for the task. The situation has been shown in Fig. 6.6, in which an instance of a task has arrived Δt time unit later than the beginning of frame k. It can be scheduled earliest in frame $(k+1)$. If the execution time of the task is more than the time difference between the start time of $(k+1)$th frame and task deadline d, the task will miss its deadline.

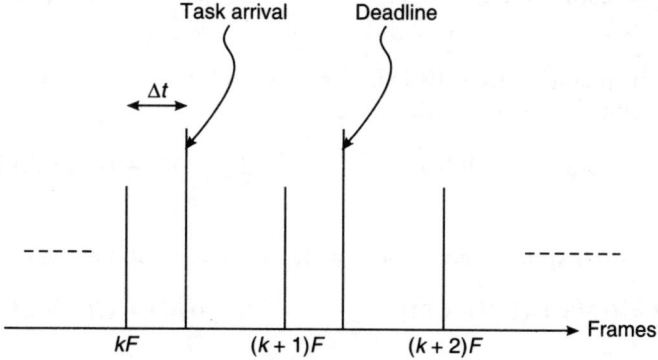

Fig. 6.6 Possibility of missing deadline.

To solve this problem, it is required that there exists at least one complete frame between the arrival of a task and its deadline. The situation has been shown in Fig. 6.7. That is,

$$2F - \Delta t \le d_i$$

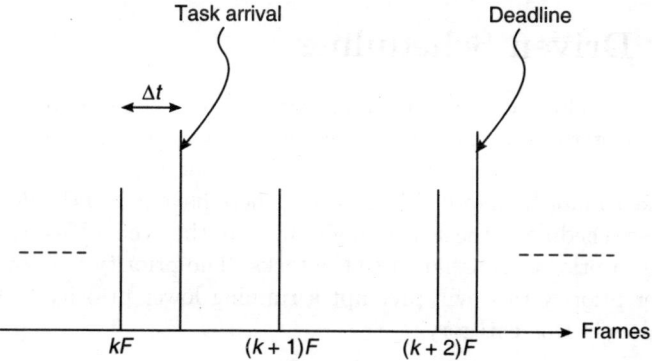

Fig. 6.7 Existence of full frame.

To refine the condition further, it can be shown that the minimum value of Δt is equal to $GCD(F, p_i)$, p_i being the periodicity of the task.

Thus,

$$2F - GCD(F, p_i) \leq d_i$$
$$\Rightarrow \qquad F \leq 0.5(d_i + GCD(F, p_i))$$

Thus, considering all tasks, the frame size has to be set.

Example 6.1

Consider a set of two tasks with deadlines 3, 4 and corresponding periodicities 5 and 6. We need to determine the optimum frame size (F) for these two tasks. First, let us try with $F = 1$. For the two tasks, it needs to satisfy the following two inequalities:

$$1 \leq 0.5(3 + GCD(1, 5)) \qquad\qquad 1 \leq 0.5(4 + GCD(1, 6))$$
$$\leq 2 \qquad\qquad\qquad\qquad\qquad \leq 2.5$$

Since both the inequalities are satisfied, $F = 1$ is a valid choice. As we need to make frame size as large as possible, let us try with $F = 2$.

$$2 \leq 0.5(3 + GCD(2, 5)) \qquad\qquad 2 \leq 0.5(4 + GCD(2, 6))$$
$$\leq 2 \qquad\qquad\qquad\qquad\qquad \leq 3$$

Thus, $F = 2$ is also a valid choice. For $F = 3$, we have the following inequalities.

$$3 \leq 0.5(3 + GCD(3, 5)) \qquad\qquad 3 \leq 0.5(4 + GCD(3, 6))$$
$$\leq 2 \qquad\qquad\qquad\qquad\qquad \leq 3.5$$

Thus, $F = 3$ cannot be used. Hence, the maximum frame size is 2 time units.

Cyclic schedulers are among the simple schedulers used in small embedded systems. The schedules are precomputed, thus allowing the design of the scheduler to be simple—a simple table look-up suffices. The table size is also reduced by introducing the concept of frames. This low memory requirement is an advantage to small systems having small memory.

6.6   Event Driven Scheduling

A basic problem with clock driven scheduling strategies is their inability to handle a large number of tasks. Determining the proper frame size becomes difficult. Moreover, in each frame, some time is wasted to run the scheduler. Other types of tasks, that is, sporadic and aperiodic tasks cannot be handled efficiently. These have led to the design of event driven schedulers. In these schedulers, the scheduling points are the events like arrival and completion of tasks. Normally, a priority is assigned to the tasks. The priority may be static or dynamic. Arrival of a higher priority task will preempt a running lower priority task. There are three important schedulers in this category.

- Foreground–background scheduling
- Rate monotonic scheduling (RMS)
- Earliest deadline first (EDF) scheduling

6.6.1 Foreground–background Scheduling

This is perhaps the simplest possible priority-driven preemptive scheduling strategy. In this approach, the periodic real-time tasks of an application have higher priority than the sporadic and aperiodic tasks. The periodic tasks run in the foreground. A background task can run only when no foreground tasks are running. There may also be multiple priority levels for the foreground tasks. At any scheduling point (dictated by arrival or completion of a task), the highest priority pending task is taken up for execution.

6.6.2 Rate Monotonic Scheduling

Rate monotonic scheduling (RMS) is a static-priority-based event-driven scheduling algorithm for periodic tasks. The priorities are assigned to the tasks based on their periodicity values. A task with lower periodicity value (that is, more number of occurrences within a fixed time interval) is assigned higher priority. Just like other preemptive priority-based schemes, arrival of a higher priority task will force preemption of any running lower priority task.

The theory underlying RMS is the *rate monotonic analysis* (RMA), that uses simple model of the system stated as follows:

1. All processes run on a single CPU, thus there is no task parallelism.
2. Context switching time ignored.
3. Execution time for different invocations of a task are same and constant.
4. Tasks are totally independent of each other.
5. The deadline of an instance of a task occurs at the end of its period.

The theory has been stated in the form of a theorem as follows:

Given a set of periodic tasks to be scheduled using a preemptive priority scheduling, assigning priorities, such that the tasks with shorter periods have higher priorities, yields an optimal scheduling algorithm.

It may be noted that the term *rate monotonic* refers to the fact that monotonically higher priorities are assigned to tasks having higher rates of occurrences. The optimality criteria states that if a schedule meeting all the deadlines exists with fixed priorities, then the rate-monotonic approach will definitely identify a feasible schedule.

Let us consider an example to understand the RMS principle. Table 6.3 shows a set of three tasks along with their execution times (e_i) and periodicities (p_i). The task $P1$ has the smallest period, thus it has the highest priority followed by the tasks $P2$ and $P3$.

Table 6.3 Example of RMS

Task	Execution time (e_i)	Period (p_i)
$P1$	1	3
$P2$	2	5
$P3$	3	15

A complete schedule for this set of tasks has been shown in Fig. 6.8. At time instant 0, it is assumed that all the three tasks $P1, P2$ and $P3$ have arrived. Further instances of $P1$ will come at time instants 3, 6, 9, 12, etc. Similarly, occurrences of instances of other tasks are also shown in Fig. 6.8. At time instant 0, since $P1$ is the highest priority task, it gets scheduled. Once it is over at time instant 1, $P2$ gets scheduled. It executes for full 2 time units to finish off. At time instant 3, another instance of $P1$ has arrived. Though $P3$ is waiting, it does not get a chance to execute as $P1$ is of higher priority. At time instant 4, none of $P1$ and $P2$ is pending. Thus, $P3$ is scheduled for a single time unit. At time instant 5, another instance of $P2$ arrives and preempts the running lower priority task $P3$. The process continues and at time instant 14, no task is ready for execution. Thus, the CPU remains idle. Hence, this set of tasks can be scheduled in RMS principle.

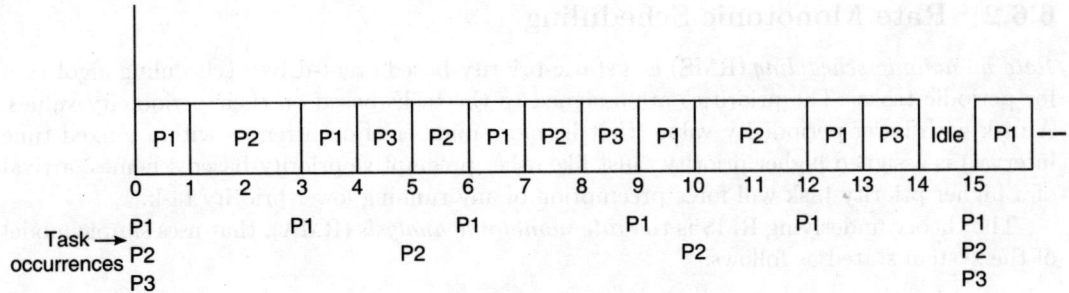

Fig. 6.8 RMS scheduling.

Next, we concentrate on a set of results that answers the question that whether a given set of periodic tasks can be scheduled using the RMS technique and ensure that none of them will miss their deadlines.

1. *Necessary condition:* For a set of periodic tasks to be scheduled using RMS principle, the sum of CPU utilizations of individual tasks be less than or equal to 1. That is,

$$\sum_{i=1}^{n}(e_i/p_i) = \sum_{i=1}^{n} u_i \le 1$$

   where, e_i is the worst case execution time of the task and p_i is its periodicity. For our example, considered previously,

$$\frac{1}{3} + \frac{2}{5} + \frac{3}{15} = 0.93 \le 1$$

   Thus, the tasks satisfy the necessary condition.

2. *Sufficient condition: Liu* and *Leland* derived a sufficiency condition to ensure that a set of tasks is schedulable using RMS. It says that a set of n real-time periodic tasks is schedulable if,

$$\sum_{i=1}^{n} u_i \le n(2^{1/n} - 1)$$

   where, u_i is the CPU utilization of task i as calculated earlier. Setting the value of $n = 3$, we get that for 3 tasks to be RMS schedulable, total utilization be $\le 3(2^{1/3} - 1)$, that is, 0.78. However, for the set of tasks in our example, the total CPU utilization is $0.93 > 0.78$. Thus, our example fails the sufficiency test, though it is still RMS schedulable.

As per the sufficiency condition, the maximum achievable CPU utilization is given by,

$$\lim_{n \to \infty} n(2^{1/n} - 1)$$

By applying L'Hospital's rule, it can be verified that the value of the expression is 0.692. However, it is not uncommon to have a set of tasks that fails the sufficiency test, but is still schedulable using RMS technique. Satisfaction of sufficiency condition only guarantees the rate monotonic schedulability.

Another test, known as *Lehoczky test* has been introduced to check whether a set of tasks is RMS schedulable or not. It can be stated as follows:

A set of periodic real-time tasks are schedulable using RMS technique under any task phasing if all the tasks meet their respective first deadlines under zero phasing.

A formal proof of the above statement is beyond the scope of this book. Interested readers may refer to books on *Operating Systems*. However, an intuitive justification can be achieved by considering a definition of *critical instant*. For a task P_i, its critical instant is the point at which an instance of the task arrives but has to wait for the longest period of time, as all other higher priority tasks are also ready at that point. This low priority task can be scheduled only at a time when no other higher priority tasks are pending. It is obvious that more number of instances of higher priority tasks will occur during a period of lower priority task, when these tasks are in-phase with each other, rather than being out of phase. Hence, if with zero phasing a task meets its deadline, it will meet all its deadlines with non-zero phasings also.

Let a set of tasks $P_1, P_2, \ldots, P_n$ with corresponding execution times $e_1, e_2, \ldots, e_n$, and periodicity values $p_1, p_2, \ldots, p_n$ be given. We have to check whether they pass the *Lehoczky test* or not. Without any loss of generality, we assume that the tasks have been ordered in their decreasing order of priority. That is, P_1 is of highest priority, while P_n has the lowest priority. Now, the task P_i meets its first deadline only if,

$$e_i + \sum_{k=1}^{i-1} (p_i/p_k) \times e_k \leq p_i$$

This is because within the period p_i of task P_i, a higher priority task P_k can appear (p_i/p_k) times. All these instances are to be scheduled before the first instance of task P_i gets scheduled. Thus, checking for all the tasks in the system in this manner, if all of them can be scheduled before their first deadlines, the set is schedulable using RMS principle.

In our previous example, for task P_1, $e_1 = 1 \leq p_1 = 3$.

For task P_2,

$$e_2 + (p_2/p_1) \times e_1 = 2 + (5/3) \times 1 = 2.6 \leq p_2 = 5.$$

For task P_3,

$$e_3 + (p_3/p_1) \times e_1 + (p_3/p_2) \times e_2 = 3 + (15/3) \times 1 + (15/5) \times 2 = 14 \leq p_3 = 15.$$

Thus, all tasks pass the test. Hence, the set of tasks is schedulable using RMS technique.

It should also be noted that a set of tasks, if not satisfying the *Lehoczky's test*, may still be schedulable using RMS policy. This can happen because the test takes all tasks to have zero phasing. In reality, if phases of tasks are different, there may be some free time for the CPU in which it can be assigned some task so that the deadlines are met.

The major advantage of RMS is its simplicity. The scheduler simply gets invoked at every interrupt from the timer, scans the list of ready tasks in their priority order and schedules the highest priority ready task. Since the priorities are static, a simple array of tasks can be used as the data structure. Almost all the commercial real-time operating systems support RMS policy. Another good feature of RMS is its ability to handle *transient overloads* efficiently. If a lower priority task's execution time gets extended temporarily, it cannot stop a higher priority process from getting scheduled. This happens because as soon as a higher priority task arrives, the lower priority one is preempted. Thus, it cannot lead to higher priority tasks missing deadlines.

The disadvantages of using RMS are the following:

1. It is difficult to handle aperiodic and sporadic tasks. This happens because their periodicity values are not known, thus no priority assignment can be done for them.

2. RMS does not produce optimal results if the deadlines differ from periodicity of tasks. In this direction, there exists another algorithm, called *Deadline Monotonic Algorithm* (DMA). In DMA, priorities are assigned to the processes based on their deadlines. Tasks with shorter deadlines will have higher priorities. Naturally, if deadlines are equal to periodicity values or are proportional to those, RMS and DMA yield same results. However, if deadlines are different from periodicity, there may be cases in which RMS fails, but DMA can produce valid schedules. On the contrary, whereever DMA fails, RMA also fails.

Context switching overhead

In our discussion about the RMS policy, so far we have ignored the context switching time. However, this needs to be considered, as a large number of context switchings can result in considerable wastage of time. In general, all these context switchings may be added to determine whether a set of tasks still remains RMS schedulable. For example, for the set of three tasks that we are considering, if we include a context switching time of 0.1 unit, there are 13 context switchings leading to an overhead of 1.3 time units. In this case, the schedule is no more a feasible one, as the total time required is 15.3 time units.

Critical tasks with long periods

In RMS, priorities are assigned to the tasks based upon their periodicity values. This may some times be counterproductive. Particularly, a task having a large periodicity value is assigned a low priority. However, it may be the case that the task is very critical, and its deadline is quite close. Presence of other higher priority tasks (that is, tasks with lower periodicity values) will preclude this critical task from getting scheduled. In such a situation, a deadlock monotonic algorithm (DMA) is one solution. Other solution is to virtually divide the task into k number of subtasks. Each subtask now has the periodicity reduced by a factor k. This may increase the priority level of the task compared to others, and thus ensure that when the task arrives, it preempts any of the running tasks. It should be noted that this splitting is only done virtually to increase the priority levels. The task is not divided physically into a number of subtasks—it still executes as a monolithic task. However, there may be some problem with the schedulability test. Each virtual subtask shows a periodicity p_i/k and execution time e_i/k. Thus, the sum of utilizations from all these subtasks is equal to $k(e_i/k)(p_i/k) = k(e_i/p_i)$, whereas for the actual task, utilization is e_i/p_i. This may indicate wrongly that the set of tasks is not schedulable under RMS technique, whereas, in reality, it may be possible to obtain such a schedule.

Aperiodic tasks

Aperiodic tasks are part and parcel of any real-time system. These often correspond to the critical emergency situations which should be responded to and completed within a limited time frame. On the other hand, since these are aperiodic, we cannot assign them any static priority to be used for RMS. Thus, special care needs to be taken to handle such tasks.

Typical solution to this problem involves dedicating one (or more) periodic task(s) to *pick up* aperiodic tasks that may need to be executed. Such a periodic task is called *aperiodic server*. There are two types of aperiodic servers:

- Deferrable server
- Sporadic server

Both the servers work on a fixed *budget* for resource utilization. If a new aperiodic task arrives and there is enough budget, the task will get scheduled immediately. However, if the budget is not enough, the task has to wait till enough budget is accumulated. In the case of *deferrable server*, the budget is replenished to its full value at regular intervals of time (dictated by the periodicity of the server). This makes the implementation of the server easier, however, at the same time makes the schedulability analysis very hard due to a phenomenon called *deferred execution effect* or *jitter effect*. In particular, assume that an aperiodic task arrives near the end of the server's period. If the server has enough budget, it will be scheduled immediately. The task may utilize its budget just in time for a new server's period. Since the budget gets replenished, the task will grab the resource for another execution. Moreover, if for a large time no aperiodic task arrives, the budget will go on accumulating. It may result in admitting a large number of aperiodic tasks in a burst.

On the other hand, in a *sporadic server*, the time of budget replenishment depends on the last time at which previous aperiodic task finished. As soon as the budget is utilized, the system initiates a timer. The budget is replenished at the expiry of the timer. Thus, there is a guaranteed minimum separation between two instances of a task. This aids in considering the aperiodic tasks also as periodic ones for RMS analysis purpose.

Limited priority levels

In most of the processors and operating systems, the number of different priority levels is limited. This can create problem in handling a large number of tasks. This is because even if based upon their periodicity values, the tasks are assigned different priorities, in actual implementation, so many priority levels may not be available. This will necessitate grouping of tasks with different priorities into same level. There are several schemes for grouping tasks into priority levels. These are:

- Uniform
- Arithmetic
- Geometric
- Logarithmic

In *uniform* assignment, equal number of tasks are assigned to all the levels. In case it is not possible to distribute evenly, more number of tasks are allocated to lower priority levels. For example, if there are 11 tasks of different priorities to be assigned to 4 priority levels, the following is the distribution. Two tasks are assigned to the highest priority level. Now, 3 levels are left with 9 tasks. Thus, three tasks are assigned to each of these levels.

In *arithmetic* assignment scheme, the number of tasks assigned to successive levels form an arithmetic progression. For example, if there are 15 tasks and 5 priority levels, the tasks assigned to successive levels from the highest to the lowest level may be 1, 2, 3, 4, 5.

In *geometric* scheme, tasks assigned to successive levels form a geometric progression. For example, if there are 15 tasks to be distributed between 4 priority levels, number of tasks alloted to different levels from the highest to the lowest level may be 1, 2, 4, 8.

In *logarithmic* priority assignment process, the range of task periods are divided into logarithmic intervals. If p_{min} be the smallest period and p_{max} be the largest period, then we calculate a factor $r = (p_{max}/p_{min})^{1/n}$, n being the total number of priority levels. Now, all tasks with periods less than $p_{min} \times r$ are assigned to the highest priority level, tasks with periods between $p_{min} \times r$ and $p_{min} \times r^2$ are assigned to the next priority level, and so on. For example, consider a set of 11 tasks with periods 5, 6, 7, 8, 9, 10, 11, 12, 13, 14, 15 and the number of available priority levels to be 4. Thus, $p_{min} = 5, p_{max} = 15, n = 4$. Thus,

$$p_{min} \times r = 6.58$$
$$p_{min} \times r^2 = 8.66$$
$$p_{min} \times r^3 = 11.79$$
$$p_{min} \times r^4 = 15.0$$

Thus, the highest priority level will have tasks with periods 5 and 6. Next level will have tasks with periods 7 and 8, third level 9, 10, 11, and the fourth level 12, 13, 14, 15.

6.6.3   Earliest Deadline First Scheduling

Earliest deadline first (EDF) is a dynamic priority algorithm in which priority of an instance of a task depends on its deadline. Unlike RMS, the priority of a task is not fixed for all instances. The priority changes dynamically during the life-time of the system. At any scheduling point, among all the task instances ready for execution, the one whose deadline is the earliest is picked up for scheduling. This justifies the name *earliest deadline first*. Since the scheduling decisions are taken dynamically, it may so happen that a task set misses deadline when scheduled using fixed priority approach, however, it becomes schedulable in the EDF policy. For example, consider a set of three tasks P_1, P_2, and P_3 with worst case execution times of 1, 2, and 3 units, respectively. Assume the respective periodicity values to be 4, 6, and 8. Now, in the fixed priority assignment scheme with RMS principle, P_1 will have the highest priority, followed by P_2 and P_3. The overall utilization is given by, $u = 1/4 + 2/6 + 3/8 = 0.958$, much higher than 0.692. Figure 6.9 shows a snap-shot of the schedule as produced using the fixed priority values. Here, at time instant 6, though the deadline of P_3 is the closest one, since P_2 has got higher priority, it gets scheduled. Consequently, it leads to a deadline miss for the task P_3. On the otherhand, using EDF policy, the same set of tasks can be scheduled, so that no deadline miss occurs. This is shown in Fig. 6.10.

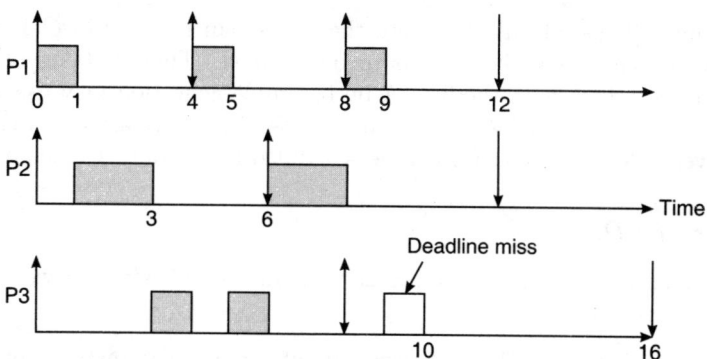

Fig. 6.9 Deadline miss with fixed priority schedule.

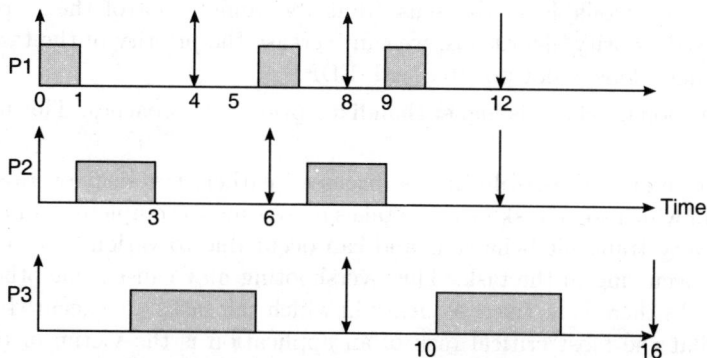

Fig. 6.10 No deadline miss with EDF schedule.

Given a set of n periodic and sporadic tasks with relative deadlines equal to their periods, the task set is schedulable by EDF if and only if the total utilization is less than or equal to unity, that is,

$$U = \sum_{i=1}^{n} e_i/p_i \leq 1$$

Thus, EDF is an optimal algorithm in the sense that if a task set is schedulable, it is schedulable by EDF as well. The phases of tasks also do not matter.

EDF enjoys the following advantages over the fixed priority-based approaches.

1. There is no need to define priorities. Based on deadlines, the priority values can change. On the other hand, for fixed-priority cases, it cannot be altered.

2. In general, EDF has lesser context switches.

3. The processor utilization is generally very high, reaching almost 100%. Thus, idle time is less.

Implementation of EDF

A very simple implementation of EDF will put all ready tasks of the system in a queue. A newly arriving task is attached at the end of the queue. For each of the tasks, their absolute deadlines are noted. Thus, an insertion into the queue can be done in $O(1)$ time. To select the next task to be scheduled, the list has to be scanned. Thus, it takes $O(n)$ time, where n is the total number of tasks in the list. A better implementation may use a priority queue as the data structure. It requires $O(\log n)$ time to insert a newly arriving task into this data structure, however, selection of the task to be scheduled next can be carried out in $O(1)$ time.

Disadvantages of EDF

EDF suffers from a number of shortcomings as well. In the following list we highlight some of these shortcomings:

1. EDF is less predictable. The response time of a particular instance of a task depends upon the deadlines of instances of other tasks. Thus, the response time is variable.

2. EDF is less controllable in the sense that we cannot control the response time of a task. In fixed-priority algorithms, we can increase the priority of the task to reduce the response time. This is not possible with EDF.

3. Implementation overhead is higher than fixed priority approaches. This has already been discussed.

4. **Transient overload problem:** As discussed earlier, transient overload refers to the situation in which some task instance takes more time to complete than the planned one. This is a very transient behaviour and can occur due to various exceptional execution sequences occurring in the task. This overshooting may cause some other tasks to miss deadlines. As there is no fixed sequence in which the tasks are executed, it is very much possible that the most critical task of an application is the victim of the least critical one overshooting its execution time estimate. It may be noted that such a situation does not occur with fixed priority approaches like RMS.

5. **Domino effect:** EDF can also show domino effect. This may occur when the utilization of a set of tasks becomes greater than 1. Here, all the tasks in the application miss their deadlines, one after the other like a set of dominos, such that, falling of one domino initiates the falling of other dominos in sequence. This has been shown in Fig. 6.11, in which there are four tasks P_1, P_2, P_3, and P_4. Their deadlines are close to each other. Total utilization for these four tasks is equal to $2/5 + 2/6 + 2/7 + 2/8 = 1.27$. As shown in the figure, all the tasks miss their deadlines. The periodicity of the tasks are assumed to be equal to their deadlines.

It may be noted that the fixed priority algorithms are more predictable. In a similar situation as in Fig. 6.12, only the lower priority tasks may miss deadlines. The tasks P_1 and P_2 never miss their deadlines. Task P_3 misses its deadline quite often, whereas the task P_4 never gets a chance to execute. That way, EDF is more *fair* as all tasks are treated in the same way!

6.7   Resource Sharing

In an embedded system, it is often required that the system resources are shared between a set of tasks. There can be two major types of resources—*preemptable* and *non-preemptable*. Preemptable resources are those which can be taken back from a task currently using it.

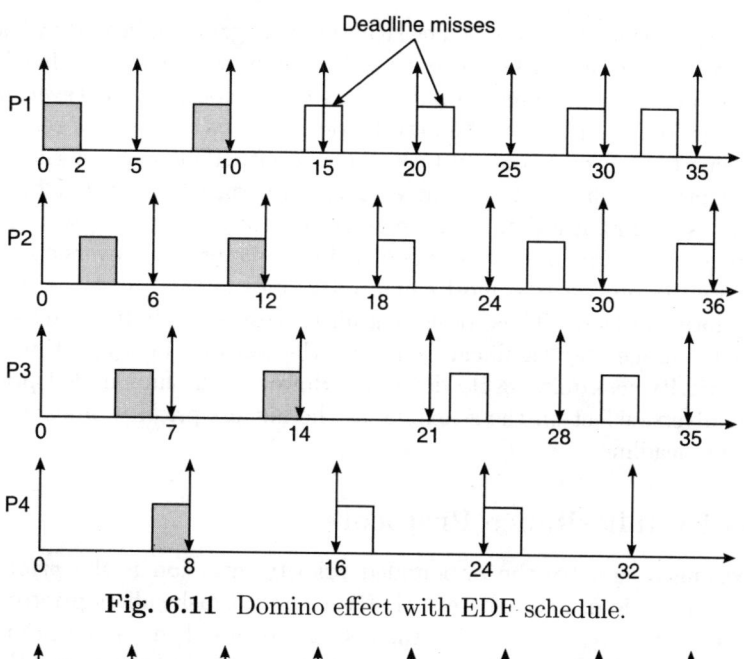

Fig. 6.11 Domino effect with EDF schedule.

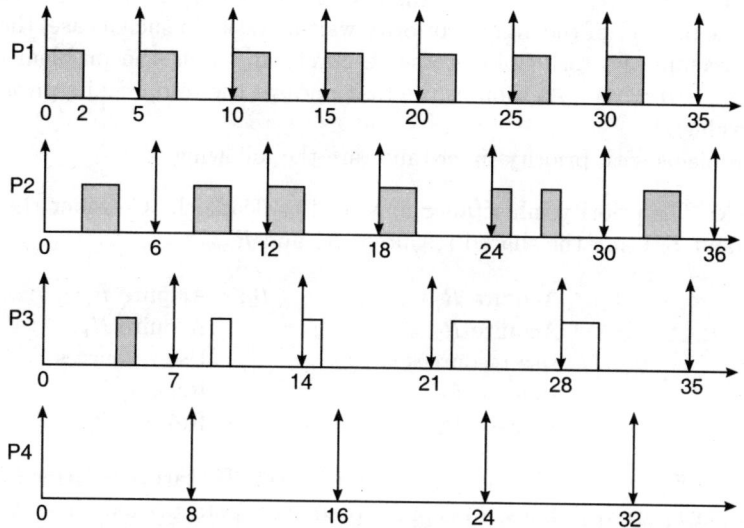

Fig. 6.12 No deadline miss for high priority tasks with fixed priority.

The task can be resumed later when the resource is again allotted to the process. Example of such a resource is the CPU. Many of the scheduling strategies that we have discussed so far are suitable for such preemptable resources only. These strategies cannot be used for non-preemptable cases as the resource cannot be taken back half-way from a process using it. Because, this may lead to an inconsistent state of the resource. Therefore, when a lower priority process has grabbed a non-preemptable resource, even a higher priority process will have to wait for the lower priority one to finish. Thus, even though the waiting process has got higher priority than the running one, it has to wait. This situation is called *priority inversion*. It is also known as *simple priority inversion*. The situation may get complicated by the introduction of some intermediate priority tasks. These tasks are not willing to use the shared resource (for which the lowest and the highest priority tasks are competing). However, due to the fact that their priorities are higher than the lowest priority task (holding the resource), they will repetitively preempt the lowest priority task. As a result, the lowest priority task cannot be completed—it continually holds the resource. This, in turn, results in indefinite wait for the highest priority task. This situation is known as *unbounded priority inversion*. It is much more severe as compared to simple priority inversions. In simple priority inversion, the delay faced by the highest priority process is at most equal to the duration for which the resource is locked by the lower priority process. Thus, designing all processes carefully to reduce the duration of critical resource usage may significantly reduce the waiting time, and thus the possibility of the highest priority task missing its deadline. However, the unbounded priority inversion may have very adverse effect on the schedule, as the highest priority process now has a high chance of missing deadline.

6.7.1   Priority Inheritance Protocol

The basic mechanism to solve the unbounded priority inversion is the *priority inheritance protocol* (PIP). The basic principle here is that whenever a task suffers priority inversion, the priority of the lower priority task holding the resource is raised through a priority inheritance mechanism to the priority of the highest priority waiting task. In such a case, the intermediary tasks cannot preempt the task holding the resource, and thus the problem of unbounded priority inversion is resolved. As soon as the task releases the resource, its priority is reverted to its original value.

Two serious problems with priority inheritance are the following:

1. **Deadlock:** The priority inheritance may lead to deadlock. Consider the following two tasks T_1 and T_2 using the shared resources R_1 and R_2.

T_1:	Acquire R_1	T_2:	Acquire R_2
	Acquire R_2		Acquire R_1
	Use resources		Use resources
	Release R_2		Release R_1
	Release R_1		Release R_2

   Assume that T_1 has higher priority than T_2, however, T_2 starts executing first. It acquires R_2 and then T_1 arrives. Thus, T_2 is preempted and T_1 starts executing. After some time, it acquires R_1 and then tries to acquire R_2 held by T_2. Now T_1 gets blocked, and due to priority inheritance, T_2 acquires the priority of T_1. It tries to acquire resource R_1 and gets blocked in the process. Thus, both the tasks T_1 and T_2 are blocked and this leads to a deadlock situation.

2. **Chain blocking:** This situation may be faced by a process trying to acquire a number of resources. It may so happen that the n resources required by the process are all held by different low priority processes. Thus, the high priority process sees n priority inversions and thus, n blockings. The situation may also occur with less than n processes. To this extreme, let us assume that T_1 is the high priority process trying to acquire the resources R_1, R_2, ..., R_n in sequence. All the resources are held by a currently running

low priority process T_2. When T_1 arrives, it preempts T_2 and tries to acquire R_1. A priority inheritance occurs and T_2 resumes execution with T_1 being blocked. As soon as T_2 releases the resource R_1, its priority gets restored to the original value. Thus, T_1 restarts, acquires R_1 and then attempts to acquire R_2 (held by T_2). Again, T_1 gets blocked and priority inheritance occurs. In this way, the situation may lead to n number of priority inversions and blocking.

To resolve the problems, some protocols have been proposed. Two important ones in this category are *Highest Locker Protocol* (HLP) and *Priority Ceiling Protocol* (PCP).

Highest Locker Protocol (HLP)

In this scheme, every critical (that is, non-preemptable shared) resource is assigned a *ceiling priority*. The ceiling priority is defined as the priority of the task having maximum priority that may request for a resource. Now, for any task acquiring the resource, its priority is set to be equal to the ceiling priority of that resource. If a task holds multiple resources, it acquires the highest ceiling priority of all the resources.

It can be shown that HLP solves the problems of unbounded priority inversions, deadlock and chain blocking. However, it introduces a new type of inversion, namely, inheritance related inversion. This occurs because as soon as a low priority task acquires a resource, its priority is increased to the ceiling value for that resource. Now, a number of intermediate-priority tasks (which may not be requiring that resource) will have to wait for the running task (which actually has lower priority) to release the resource. This is true even for the situation in which the low priority task has not yet requested to acquire the resource.

Priority Ceiling Protocol (PCP)

This approach solves the problems of unbounded priority inversion, chain blocking, and deadlock. Moreover, it reduces the possibility of inheritance related inversions. In this scheme, a task requesting a resource may not be alloted the resource, even if the resource is free. It follows a specific resource-grant rule. For this purpose, every resource R_i has a ceiling priority CR_i as in the case of HLP. There is another system-wide ceiling, known as *current system ceiling* (CSC) to keep track of the maximum ceiling value of all resources active at any instant of time. The CSC is initialized to a value lower than the priority value of the lowest priority task in the system.

1. *Resource grant rule:* It consists of two clauses to be applied when a task T_i requests for a resource.

    (a) *Resource request clause:*
        i. The resource is allocated if the task is currently holding another resource with ceiling priority equal to CSC.
        ii. Else, T_i will be granted the resource only if its priority is greater than CSC. In both the cases, if the resource is allocated to T_i and the value of CSC is less than the ceiling value of the resource, CSC is updated to this ceiling value.

    (b) *Inheritance clause:* Whenever a task is prevented from acquiring a resource by failing to meet the resource grant clause, it blocks and the task holding the resource inherits the priority of the blocked task, if its priority is lower.

2. *Resource release rule:* When a task releases a critical resource, the CSC value is updated to be the highest ceiling priority of all resources currently under use. The task releasing the resource either gets back its original priority or the highest priority of all tasks waiting for any resources which it might still be holding, whichever is higher.

It may be noted that in PCP, the priority of a task does not increase automatically on acquiring the resource—only the value of CSC changes. Only when some higher priority task gets blocked, the priority value is inherited by the lower priority process. Thus, the possibility of inheritance-related inversions gets reduced.

6.8   Other Features of RTOS

Apart from their differences on scheduling policies, an RTOS has some more distinctive characteristics compared to an ordinary operating system. Some of these are as follows:

1. **Timers:** Any real-time system should support high-precision timers, often more than one in number. The resolution is much higher than that available in ordinary OS. There are different types of timers that may be present in an RTOS.

   (a) *Periodic timers*—used normally for sampling events at regular intervals of time performing some periodic tasks.
   (b) *Aperiodic/Watchdog timers*—these are one-time timers, normally used to detect missing deadlines. Whenever a critical function is started which needs to be finished within a specific time duration, a watchdog timer is started. If the procedure finishes within the stipulated time, the timer is reset. However, if the procedure is not over, the timer expires and an exception is generated, indicating a deadline miss for the task.

2. **Task priorities:** In normal operating system, the priority of a task may change over its life-time. For example, depending upon the resource usage, the priority may increase or decrease. However, in RTOS, the tasks will have some specified priorities which do not change due to reasons other than priority inheritance. In normal OS, priorities are modified dynamically by the system to improve the system throughput. However, in real-time systems, the goal is not to optimize system throughput, but to ensure that all deadlines are met.

3. **Reduced context switching time:** Context switching time is to be reduced in RTOS, since when a critical task arrives, it should start immediately. It should not be waiting a lot simply for the running task to leave CPU. In traditional OS, the kernel is made non-preemptive, so that system calls can execute atomically. However, this is not allowable in RTOS. The kernel of a real-time operating system must be preemptable. The worst-case preemption time should be of the order of a few microseconds.

4. **Interrupt latency:** It is defined as the time required to invoke the corresponding interrupt service routine (ISR) from the point of occurrence of the interrupt. In real-time systems, it is expected that the upper bound of the interrupt latency be small, less than a few microseconds. Once an interrupt occurs, to ensure that lot of time is not spent in processing the interrupt, often a policy called *deferred procedure call* (DPC) is used. Low interrupt latency is achieved by transferring bulk of the activities of ISR to a

deferred procedure. The DPC performs most of the activities of ISR, but executes (after ISR completes) at a lower priority value.

5. **Memory management:** Efficient memory management is one of the very important issues in any operating system design. Most of the contemporary general purpose OS support the virtual memory and memory protection features. Virtual memory, though practically makes the amount of memory available to a process infinite, at the same time, it makes the execution time of the program unpredictable. This happens because in the event of a page-fault, fetching the desired page from the secondary storage incurs significant latency. Thus, generally, in RTOS, the concept of virtual memory is avoided. In systems that support virtual memory to cater to the non-real-time tasks, *memory locking* feature is provided to ensure predictable execution of real-time tasks. Memory locking prevents a page from being swapped from memory to hard disk.

In the absence of virtual memory, in RTOS, memory is often allocated as chunks to the tasks. This may lead to problem of *memory fragmentation* (creation of large number of small free chunks). Once the memory gets fragmented, it becomes very difficult to allocate memory to new requests. Moreover, protection also becomes difficult. In virtual memory environment, protection is implemented at page level. The user pages can easily be separated from the kernel pages, and thus, modification to OS code in kernel can be controlled. However, in RTOS, due to the absence of such protection, kernel code and user code execute in the same space. Thus, a system call and a function call within an application are indistinguishable. This makes debugging of user code difficult, as a corrupted pointer may easily modify system code as well.

6.9 Commercial RTOSs

In this section we will look into a few popular commercial real-time operating systems and a comparison between them. The *IEEE Portable Operating System Interface for Computer Environments, POSIX 1003.1b* provides a set of compliance criteria regarding the services to be provided by any RTOS. This will aid in the porting of applications between RTOSs. The following are some of the important requirements.

- Asynchronous input–output
- Synchronous input–output
- Memory locking
- Semaphores
- Shared memory
- Execution scheduling
- Timers
- Inter-process communication primitives
- Real-time files
- Real-time threads

Before going into the individual RTOSs, we note the common capabilities of all these operating systems.

1. *Efficiency:* Most of these OSs are realized as *microkernels*, that is providing low-level interfaces for address space management, thread management and interprocess communication. It runs at the highest priority level. Rest of the portions of OS are realized as service providers (for example, the networking service). This reduces the amount of code in the kernel. Since the services are at user space, almost no context switching overhead is incurred to send messages to these services.

2. *Non-preemptable system calls:* Some portions of the system calls are made non-preemptive to support mutual exclusion. This makes the OS design simpler. These portions are highly optimized and are made as deterministic as possible.

3. *Prioritized scheduling:* For POSIX compliance, all RTOSs support at least 32 different priority levels. The number of levels can be as high as 512.

4. *Priority inversions:* Some sort of priority inversion protocols are supported by all RTOSs.

5. *Memory management:* Support of virtual memory is often present, however, it may not be at the level of paging.

6.9.1   General Purpose Operating Systems

Two most common general purpose operating systems are *Windows NT* and *Unix*. Both of them have quite a few features to handle real-time tasks to some extent. Table 6.4 gives a summary. Although Windows NT kernel is non-preemptable, it has certain points at which preemption is allowed. Real-time Unix also allows preemption points within system calls. Deferred procedure calls are supported in Windows NT, but not in Unix. Both Windows NT and Unix continually manipulate thread priorities to ensure fairness and better throughput of the system. However, Windows NT provides a band of interrupt priorities that cannot be altered. This band can be used for real-time tasks. There are 16 priority levels in this class, however, each thread is restricted to a subset of priorities in the range of ± 2 levels of its inital priority. Due to the deferred procedure call mechanism to handle interrupts, though Windows NT provides fast response, it is not a hard RTOS. Priority inheritance is also not supported in Windows NT. Thus, it may lead to deadlock.

Table 6.4   Real-time features of Windows NT and Unix

Real-time feature	*Windows NT*	*Unix*
Preemptive, priority-based multitasking	Yes	Yes
Deferred interrupt threads	Yes	No
Non-degrading priorities	Yes	No
Memory locks	Yes	Yes

6.9.2   Windows CE

It is a modular OS targeted towards mobile 32-bit devices with small memory. It slices CPU time into threads and provides 256 priority levels. All threads are enabled to run in kernel mode. This enhances the performance. To incoporate portability, an equipment adaptation layer isolates device dependent routines. The equipment manufacturer can specify trusted modules and processes to prevent unauthorized applications from accessing system application programming interfaces (APIs). The non-preemptable portions of the kernel (called *Kcalls*) are broken down into small sections reducing the durations of non-preemptable code. The kernel objects (such as processes, threads, semaphores, etc.) are allocated dynamically into virtual memory. While executing non-preemptive kernel code, all kernel data are moved into the physical memory to avoid translation look-aside buffer (TLB) misses.

6.9.3   LynxOS

It is POSIX compatible, multithreaded OS designed for complex real-time applications requiring fast, deterministic response. It can be used in a wide range of platforms—starting from small embedded products to very large ones. The kernel provides services like TCP/IP streams, I/O and file handling, sockets, etc. In response to an interrupt, the kernel dispatches a kernel thread which can be prioritized and scheduled as other threads. The priority of this thread is the priority of the user thread that handles the interrupting device. The mechanism ensures predictable response time, even in the presence of heavy I/O. OS depends on hardware memory management units for memory protection, however, demand paging is also supported. The scheduling policies supported are, prioritized FIFO, dynamic deadline monotonic scheduling, time-slicing, etc. It has 512 priority levels and supports remote operation.

6.9.4   VxWorks

This is another widely adopted RTOS with a visual development environment. It has more than 1800 APIs and is available on popular CPU platforms. The microkernel supports 256 priority levels, multitasking, dynamic context switching, pre-emptive and round-robin scheduling. Semaphores, mutual exclusion, and priority inheritance are also supported. It offers network support, file system and I/O management.

6.9.5   Jbed

It is a real-time OS supporting applications and device drivers written in Java. It translates byte-code into machine code prior to class loading, instead of interpreting the byte code. It supports real-time memory allocation, exception handling, automatic object destruction, etc. Hard real-time applications are supported via specific class libraries. Ten different thread priority levels and EDF scheduling policy are supported.

6.9.6   pSOS

It is an object-oriented operating system, supporting tasks, memory regions, message queues and semaphores. The scheduling strategies supported are preemptive, priority-driven, or EDF. Priority inversion is handled both via priority inheritance and priority-ceiling protocol. The application developer has full control over interrupt handling.

Table 6.5 presents a comparison between the contemporary real-time operating systems from five different angles—number of priority levels, synchronization mechanism supported, priority inversion, hosts on which it is available and kernel characteristics.

Table 6.5 Comparison between RTOSs

Name	Priority levels	Synchronization mechanism	Priority inversions	Hosts	Kernel characteristics
AMX	N/A	Mailbox, wait-wake requests	Yes	Windows	Predictable memory block availability
C Executive	32	Messages, dyanmic data queues	Yes	Windows, Solaris	
CORTEX	62	Recursive locks, mutexes	Yes, priority ceiling	Windows, Unix	CPU-independent software interrupt manager; segmented memory models
Delta OS	256	Semaphores, timers, message queues	Yes	Windows, Linux	
Ecos	1-32	Semaphores, timers and counters	Yes, priority ceiling	Windows, Linux	Soft real-time small embedded applications
Emboss	255	Mailbox, binary and counting semaphores	No	Windows, Linux	Task profiling, activation time of tasks independent of number of tasks
ERTOS	256	Thread messaging, queues, semaphores	No	Windows, DOS, OS/2	High speed interrupt driven serial port
INTEGRITY	255	Semaphores	Yes		Critical embedded applications, object oriented, task profiling, distributed processing
IRIX	255	Message queues	Yes	SGI	Double precision matrix, scalable
RT-Linux	1024	Shared memory, files	Yes, lock free data structures, priority ceiling	Linux	Hard real-time applications
QNX Neutrino	64	Message passing	Yes, priority inheritance	Windows, Solaris, Linux, Symmetrical multi-processors	OS components in MMU protected address spaces

6.10 Conclusion

In this chapter, we have seen a detailed view of the features of real-time operating systems, how do they differ from normal OS, and also the pros and cons of various techniques prevalent in RTOS. Using the facilities provided by the RTOS, the embedded system designer can select how to execute the set of processes that are needed in the realization of the application. The assignment of priorities can be guided to perform scheduling. In the next chapter, we will look into the various specification techniques that may be utilized to specify the behaviour of an embedded system, from which the design will evolve.

Exercises

6.1 How does a real-time operating system differ from an ordinary operating system? How do their performance evaluation criteria differ?

6.2 What are the different types of real-time tasks? Give a few examples of each.

6.3 Classify tasks based on their periodicity values. Give examples.

6.4 Why do you think that most of the tasks in an embedded application be periodic in nature? Mention the different quantities by which a periodic task is characterized.

6.5 Why is scheduling of tasks important in any RTOS? What are the optimization criteria on which the quality of a scheduling algorithm depends?

6.6 What do you mean by a valid schedule and a feasible schedule?

6.7 What are scheduling points? For clock-driven and event-driven schedules, what are the scheduling points?

6.8 Distinguish between table-driven and cyclic schedulers.

6.9 For the following task set, perform table-driven scheduling.

Task	Execution time	Periodicity
T1	1	5
T2	2	10
T3	4	15
T4	3	15

6.10 For the previous problem, determine the frame-size for cyclic scheduling.

6.11 Why is it mandatory that one full frame must exist between the arrival of a task instance and its deadline?

6.12 What are the advantages of event-driven scheduling over clock-driven ones? Mention different types of event-driven schedulers.

6.13 Why is RMS called a static priority algorithm whereas EDF is a dynamic priority one?

6.14 Perform RMS schedule for the set of tasks in Exercise 6.9. Perform Liu-Leland and Lehoczky tests to check whether the set of tasks is RMS schedulable or not.

6.15 Construct an example task set so that it fails Leland's test but is still RMS schedulable. Does it pass Lehoczky's test?

6.16 State the problem faced by critical tasks with long periods in RMS schedule. How is the problem solved in arriving at an acceptable schedule?

6.17 How can aperiodic tasks be handled in RMS schedule? In this context, enumerate the difference between deferrable and aperiodic servers.

6.18 What problem is posed to the RMS scheduling policy by the limited number of priority levels available with general processors? How is it resolved?

6.19 Suppose there are 100 tasks $T1$ through $T100$, with the priority of the ith task being i. Determine how the tasks will get distributed into 10 priority levels using uniform, arithmetic, geometric and logarithmic assignment schemes.

6.20 For the set of tasks in Exercise 6.9, perform an EDF scheduling.

6.21 Compare RMS and EDF scheduling based upon their relative advantages and disadvantages.

6.22 Why is it that the transient overload problem does not affect the RMS scheduling policy, whereas, it may affect EDF seriously?

6.23 Explain domino effect.

6.24 What is priority inversion? Can there be priority inversions in the following cases:

(a) A lower priority process asking for a resource currently held by a higher priority process.

(b) A higher priority process asking for a resource currently held by a lower priority process.

6.25 What is the difference between simple priority inversion and unbounded priority inversion? Give an example of each case by considering a set of periodic tasks.

6.26 Explain priority inheritance protocol. Can it solve all types of priority inversions? What are the problems introduced by the basic priority inheritance protocol?

6.27 How is priority inversion handled in highest locker protocol and priority ceiling protocol?

6.28 What are watchdog timers? How do they help in designing real-time tasks?

6.29 Why is virtual memory not a very desirable feature for real-time operating systems?

6.30 What is meant by memory locking? In which type of environment is it used?

6.31 Why is the incorporation of preemption points within real-time kernels important? Which operating systems support them?

6.32 Compare between the real-time operating systems in terms of their features.

6.33 Explore literature for details of other real-time operating systems.

CHAPTER 7

Specification Techniques

7.1 Introduction

Specification of a system is the first step towards its realization. After performing the requirement analysis, the specification process starts. Since an embedded system consists of a set of sequential/parallel subtasks, their specification should also be able to capture this. The interaction between the tasks be clearly spelt out and at the same time, the technique be easy to write, understand and verify the system. There are many specification methodologies that have been developed, but before going to these, let us have a look at the desirable features of a specification technique.

1. A specification technique should have the ability to describe a system from various angles, such as timing, state-oriented and event-oriented approaches. Timing is one of the most important issues in embedded system design. On the other hand, many of the control-dominated applications can most conveniently be specified using a state-oriented approach. Finally, the embedded systems are generally reactive in nature, responding to the events occuring in the environment. Thus, an event-oriented approach is also a desirable feature of the specification technique.

2. Support for hierarchy is another very important feature. In the absence of hierarchy, the specification may become too big to handle either manually or electronically. Thus, usage of hierarchy, concealing the details at higher levels, can effectively make the specification more amenable for understanding and design.

3. Concurrency, synchronization and communication are the three essential aspects of distributed embedded systems. The specification technique must support these aspects efficiently.

4. It is very much essential to verify an embedded system specification for its correctness. For doing this, it is desirable that the specification be executable. Thus, a technique based on some programming language appears to be a better choice from this perspective.

5. Automated synthesis is another important desirable feature of embedded system specification. This means full- or semi-automatic tools or techniques to convert a specification into a design.

6. Feature should also be present to specify non-functional quantities, like size, weight, fault-tolerance, user-friendliness, power consumption, life-time, etc., of the system to be designed.

The specification techniques reported in the literature may have one/more features missing from this list. The techniques are built around some *models of computation* (MOC). One of the very important MOCs is due to *Von Neumann*. However, this model does not have any notion of time built with it. Thus, it is not much suitable for modelling embedded systems. For this, many other models have been developed that can effectively capture timing and concurrency. The following are some of those models:

- Communicating finite state machines (CFSM)—like StateChart, SDL, etc.
- Discrete event models like VHDL, Verilog, etc.
- Asynchronous message passing
- Synchronous message passing

In subsequent sections we will discuss about some of these important techniques.

7.2   StateChart

StateCharts are among the widely used specification techniques, suitable particularly for control-dominated embedded system specifications. It is a modification of *finite state machine* (FSM) commonly used in controller specification. It may be noted that the normal FSM specification suffers from the following drawbacks:

1. Complexity of the state diagram increases dramatically as the number of possible states increases. This poses a major bottleneck in modelling large systems.
2. In traditional state machines, there is no concept of hierarchy. This, in turn, leads to the previous problem of state-explosion.
3. There is no support for concurrent constructs. Traditional state machine modelling is based on sequential transitions from one state to the next. Concurrent systems cannot be modelled in this manner, as various portions of the system may be in different states.

StateCharts introduced by *David Harel* in 1987, have some special features to overcome these problems. It allows the system to be simultaneously in more than one state. The concurrency can be modelled easily. The concept will become more clear when we discuss about the *superstates*. The states can communicate and synchronize between themselves. Some of the important features are as follows:

1. All orthogonal regions of the chart (that is, different subsystems within the whole system) accept events sent to the object.
2. One region may create an event as a result of a transition that is consumed by another orthogonal region.
3. A guard may be used to test if another region is in a certain state before allowing a transition to occur to the guarded state.

7.2.1   Modelling Hierarchy

System hierarchy can be modelled in StateCharts using the concept of *superstates*. A superstate contains other states within it. These are called substates of the superstate. The substates may again be superstates. A state which is not composed of any other state is called a *basic state*. It may be noted that a state may belong to more than one superstate.

There can be two types of superstates:

1. **OR-superstate:** This is commonly available in FSMs. The system, at any point of time, is in one of the constituent substates of an OR-superstate. For example, in the full FSM, it is always in one of its substates.

2. **AND-superstate:** An AND-superstate encompasses a number of substates, the system can be in all these states simultaneously. Though this apparently seems counter-intuitive, it can be utilized to model systems with different subsystems. For example, as shown in Fig. 7.1, the state S is an AND-superstate. It has two substates, each of which is an OR-superstate. Thus, the system can be simultaneously in one of the basic states $\{s_1, s_2, s_3, ...\}$ for the first subsystem, and in one of $\{p_1, p_2, p_3, ...\}$ for the second subsystem. If the number of states in the first subsystem is n_1 and that in the second subsystem be n_2, then the total number of states needed in the AND-superstate is n_1+n_2. However, in the absence of such a feature, a total of $n_1 \times n_2$ states will be needed to represent the system S. The AND-superstates are also said to consist of *orthogonal states*. Each of these orthogonal states can be used to model different regions of the system. For example, a telephone handset has at least two different functions—line monitoring and user interaction through buttons. The individual functions have their own state-transition behaviours. Thus, at any point of time, each of the subsystems is in one state. The overall handset can be considered simultaneously in two states. AND superstate helps in modelling this type of systems.

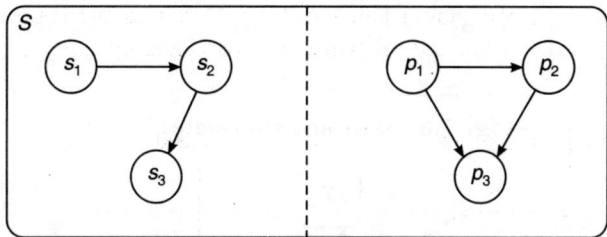

Fig. 7.1 AND-superstate.

Next, we will highlight some of the features of *StateChart*.

1. *Transitions from superstates:* Figure 7.2 shows a hierarchical StateChart in which the state S is an OR-superstate. While the system is in state S, it is in either of the states P or Q. From state P, on occurrence of event C, it transitions to state Q. In whatever substate the system be, on occurrence of event E, it will switch from state S to state T.

2. *Start and stop states:* Like FSM, StateChart may also have explicit start and stop states. Start states are indicated by simple large dots, whereas the stop states are encircled dots. For example, as shown in Fig. 7.3, the overall StateChart has S as the start state. Within S, P is the start state. That is, whenever the state S is entered, unless otherwise stated, the state P will be reached. Thus P is also called the *default state* of S.

3. *History mechanism:* Sometimes, it may be required that the system when transiting into a superstate, should enter into the substate it was in the last time. This can be accomplished by the history mechanism (as shown in Fig. 7.4). Here, while transiting out of state S, the system remembers the substate in which it was. Later on, while

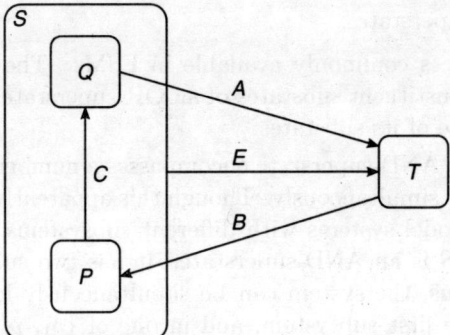

Fig. 7.2 Transitions from superstate.

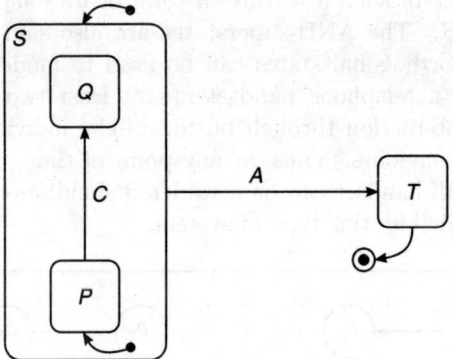

Fig. 7.3 Start and stop states.

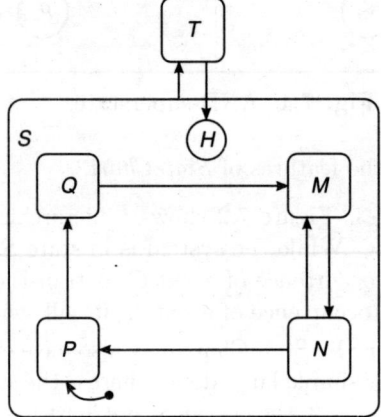

Fig. 7.4 History mechanism.

transiting from T to S, the arrow reaching H indicates that it should restart from the state the system remembered.

4. *Timers:* These are very important for real-time embedded systems. A timer is also represented as a state with a time value. Once the state has been entered, if no event

occurs to create a transition out of the state, the timer will time out and the transition marked *timeout* from this state will take place. This has been shown in Fig. 7.5, in which the state P is a timer with a timeout period of 10 ms. If the event x does not occur within this 10 ms time, the transition labeled *timeout* to state S takes place.

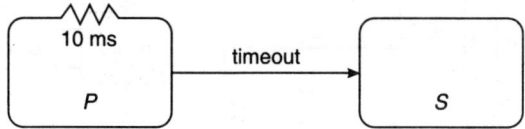

Fig. 7.5 Timer state.

5. *Edge labels:* The transitions in a StateChart are labeled as,

$< event > [< condition >][/ < reaction>]$

Thus, there must be an *event* associated with a transition. The $< condition >$ acts as a Boolean guard. It is an expression involving values of variables. The *condition* is enclosed within '['and']'. The $< reaction >$ part sets some variables to new values. Thus, some typical edge labels can be,

$$key\text{-}pressed \ / \ on := 1$$
$$key\text{-}pressed \ [on = 1]/ \ on := 0$$
$$key\text{-}pressed \ [on = 0]/ \ on := 1; \ light = 1$$

6. *StateChart simulation:* StateChart execution consists of three phases. In the first phase, the events and conditions are evaluated from the current state of the system. The transitions that are enabled by the process are identified. In the second phase, for the enabled transitions, the *reactions* are evaluated, but are not immediately assigned to the variables, rather, they are kept in temporaries. In the third and final step, the transition to the new state takes place and the variables are assigned the values evaluated in the temporaries. This three-step semantics helps in simulating situations in which a variable is being modified in a transition, whereas, in another simultaneous transition, the variable appears in the *condition* part.

It may be noted that the visibility of events is limited to the step following the one in which they are generated. The values do not live forever. On the other hand, a variable, once assigned a value, will remember it till a new assignment overwrites it.

Example 7.1
Digital watch: Next, we look into the StateChart specification of a digital watch. It has three different modes—*normal display* (default), *update* to update time and date settings, *alarm-set* to set alarm. In *normal display* mode, normally the time is displayed. However, on pressing button a, it switches to display date. On further pressing of a, it comes back to display time again. From normal display mode, on pressing button c, it switches to the mode to *update* time and date settings. Similarly, on button b, it switches from normal display mode to *alarm-set* mode, in which, it can set alarm time and chime. This has been shown in Fig. 7.6. Apart from this main function, the clock has an alarm mode to ring alarms. There is a background light that can be turned on/off anytime. This has been shown in Fig. 7.7.

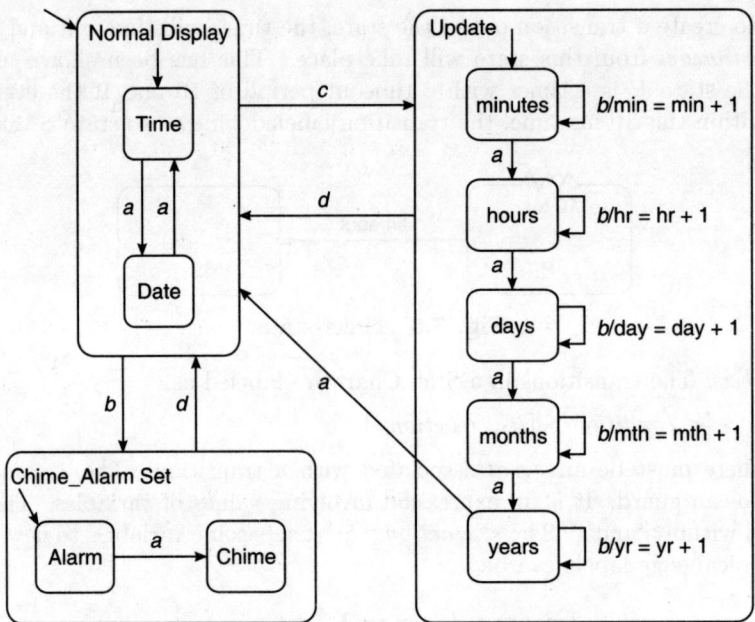

Fig. 7.6 StateChart for modes of digital watch.

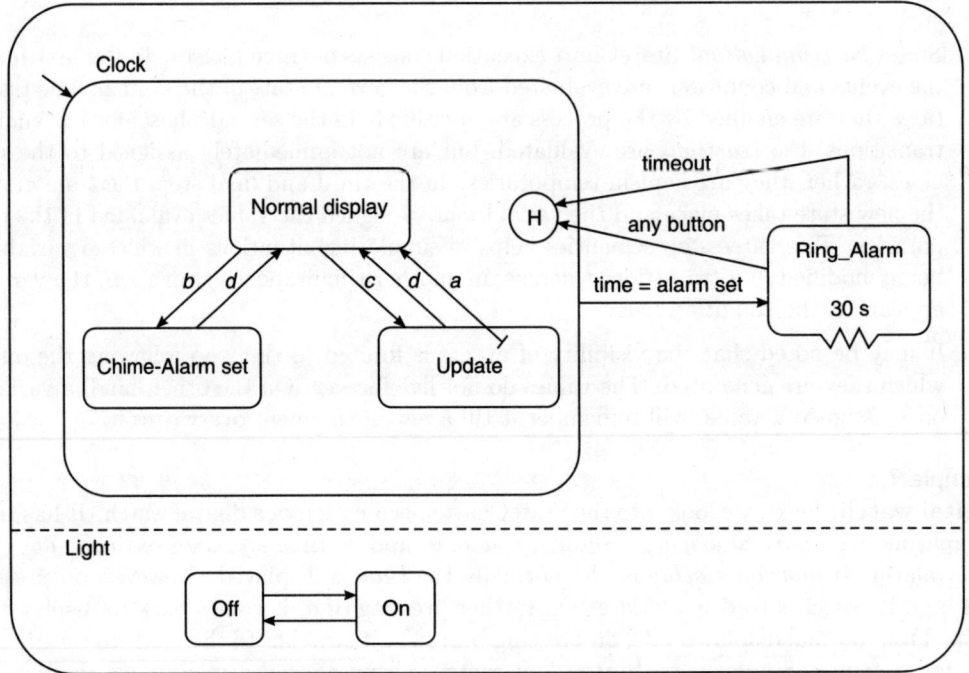

Fig. 7.7 Concurrency in digital watch.

Example 7.2
Date entry mechanism: Let us assume a system to be consisting of a numeric key-pad (keys 0 to 9) and the following special keys to enter a valid date.

1. A *delete* key to delete the character entered last.
2. A *delimiter* key to give delimiter between the day, month, and year parts.
3. An *enter* key to complete the data entry.

The corresponding StateChart has been shown in Fig. 7.8. The diagram is self-explanatory.

Disadvantages of StateChart

The following are some of the short comings identified for the StateCharts:

1. The concept of AND-superstates works well for simulation. However, if implemented in software, there hardly exists any efficient realization of those, particularly for small processors unable to support concurrent processes as in UNIX. Hardware being inherently parallel in nature, it follows the AND superstate behaviour more closely.

2. There are no object-oriented features. Thus, the object-oriented design methodologies cannot be practiced.

3. The events and variables are known to all the states in a StateChart. This requires the ability to broadcast any change in their values to the entire StateChart. This may be possible for centralized systems, but becomes a difficulty for modelling distributed systems. It may be noted that distributed systems normally work through a message-passing mechanism, in which messages are often directed to a single or a group of processes. Thus, the occurrence of an event or changing the value of a variable will create a large number of messages in the system.

4. StateCharts do not have any programming constructs. This makes it unsuitable for complex computation. It also cannot describe hardware structures and non-functional behaviour.

Thus, StateCharts are particularly suitable for centralized, control-dominated system specification. There exists a number of commercial tools like *StateMate, StateFlow, BetterState* etc. Many of these tools can automatically synthesize the corresponding C/VHDL code from the specification.

7.3   Specification and Description Language (SDL)

SDL is a premier language for specification, design, and development of real-time systems, in particular, the telecommunication applications. The currently available version is *SDL-2000*, proposed in 1999 by ITU-T and some modifications later on. It is a graphical language and is based on the concept of *Extended Finite State Machine* (EFSM). An SDL system consists of one or more communicating *agents*, with the outermost agent communicating with the environment. An agent may in turn contain other agents in a hierarchy, definition of behaviour by EFSM, data variables of value or references data types and communication-based on asynchronous message exchange.

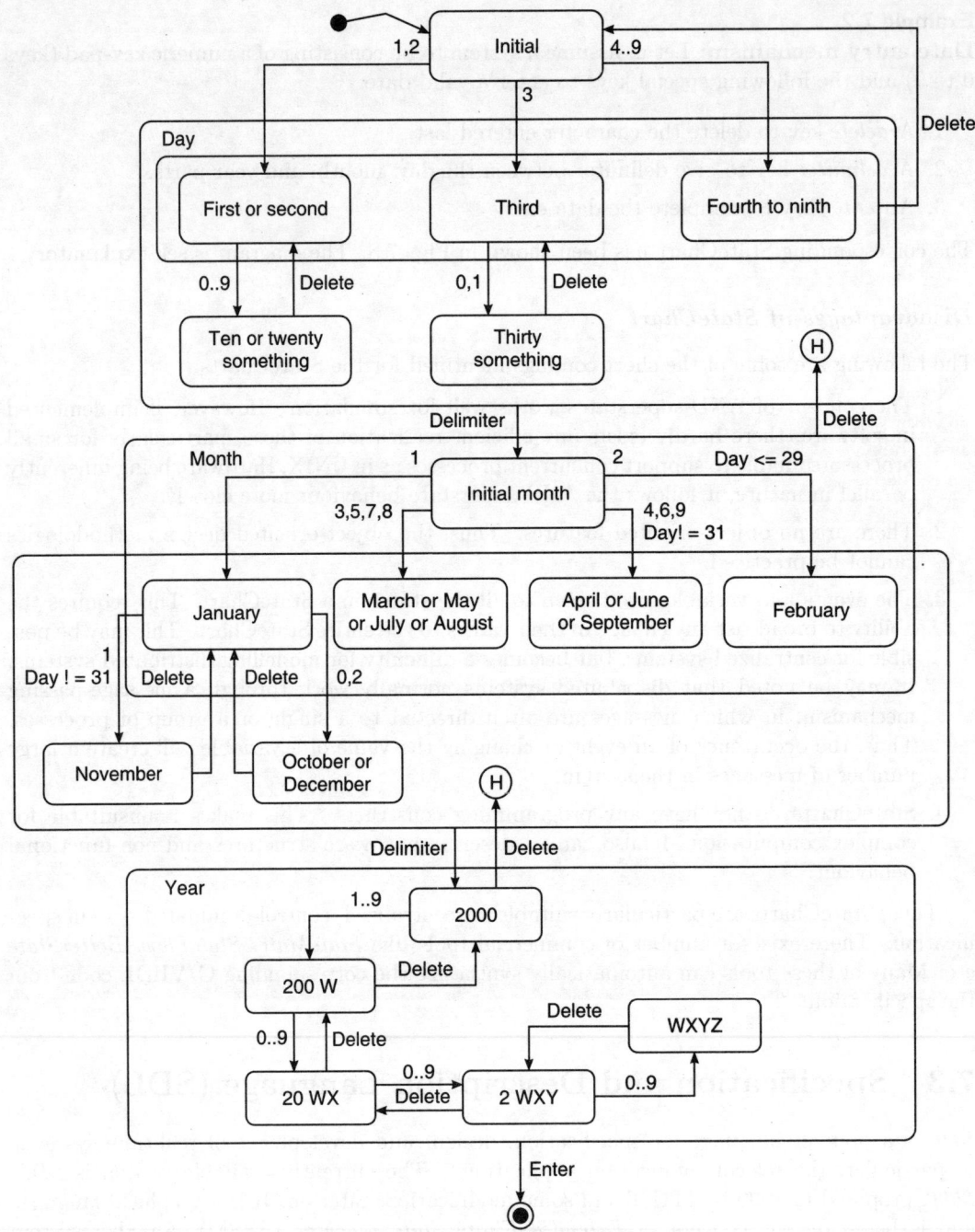

Fig. 7.8 StateChart for date entry.

SDL specification consists of a set of diagrams. The top-level diagram is the outermost agent, it shows the connection between the system components and the environment. Each diagram has one or more pages, each page having the following:

- a *frame* with some information attached to the outside.
- the *diagram heading* showing the kind and identity of the item described, in the top left corner.
- *page name* and *number of pages* in the top right corner.

The symbols normally used in a diagram are as shown in Fig. 7.9. Figure 7.10 shows a simple system-level diagram. This is a bit-stuffing system that does not allow the occurrence of bit-string "011110". For doing this, as soon as the *sender* finds a sequence "01111", it forcefully stuffs a '1' bit. In the *receiver*, this extra stuffed '1' bit is removed. The diagram consists of two processes, *sender* and *receiver*. A *channel* ($\rightarrow$) shows the signals between two agents, or between an agent and the environment of a diagram. The signal names are listed near the arrow-head between '[' and ']' symbols. The arrow-head shows the direction. Channels may also be named. Name of channels, signals, etc. can have letters, digits and underscore characters. For example in Fig. 7.10, the signals are named 0 and 1.

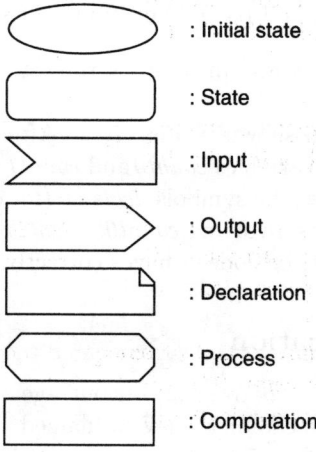

Fig. 7.9 Symbols in SDL.

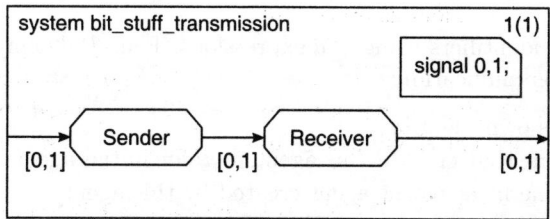

Fig. 7.10 SDL system for bit-stuffed transmission.

A system diagram can contain *process* agents or *block* agents. Instances within a block agent work concurrently and asynchronously, whereas instances within a process are scheduled one at a time. A block agent can contain within it other block agents or processes, whereas, a process can contain other process agents only within it. The boundary of agent diagram also serves the purpose of information hiding. Items defined within enclosing agent diagram are visible in inner diagrams, but not the outer way. This is different from *StateChart* that follows a broadcast mechanism.

An agent containing a single state machine can have a behaviour graph as its diagram. Figure 7.11 shows the *sender* process containing a finite state machine. It has a *start* state and other states, like *initial, 0, 01, 0111, 011111* along with associated inputs and outputs. From the initial state onwards, based upon the input bits, it traverses through the states. For example, from the state *initial*, upon getting an input 0, it outputs a bit 0 and goes to state 0. While, on getting 1, it outputs 1 and loops back to the state *initial*. Finally, from state 0111, upon getting another 1 as input, it outputs two 1's and goes to a state named 011111. Thus, it pads one extra 1 to avoid the pattern "011110" in the output.

SDL extends the concept of FSM in the following two ways.

1. Each agent has got a queue associated with it that stores all the signals received in a first-in-first-out manner.

2. Data can be received in signals, stored in variables, used in expressions, used to decide the behaviour of the agent, and passed out in output signals.

The FSM waits in its current state until one of the signals that can be consumed in that state is available.

The *receiver* process has been shown in Fig. 7.12. After start, it waits in state *wait* and scans the queue for a 0 or 1. Two variables *count0* and *count1* have been declared and initialized in the data declaration *dcl*. It uses the symbols *decision* (for example, checking *count0, count1* etc.), *tasks* (such as, computations involving *count0, count1*), *text* (the *dcl*), and *stop* (×). The stop condition is not reachable if the code behaves correctly.

7.3.1 Signal Communication

Signals are transmitted via *channels* from an output agent to an input agent. Signals are declared just like variables. The channels may be named. For example, consider a process interaction diagram as shown in Fig. 7.13. Here, block $B1$ has two processes $P1$ and $P2$. The signals A and B can be sent over the channels $Sw1$, $Sw2$, or $Sw3$. Now, in the behaviour of the processes, the signal transmissions can be specified in different ways as follows:

1. Through process identifiers using *Pid* expressions. Four *Pid* expressions are available to each agent for communication:

   - *self*—agent's own identity
   - *parent*—agent that created this agent, *null* for initial agents
   - *offspring*—the most recent agent created by this agent
   - *sender*—the agent that sent the last signal input

2. Through explicit channel names, such as $Sw1$, $Sw2$, $Sw3$, etc.

3. Implicitly, by the fact that the signal name identifies the channel as well. For example, in Fig. 7.13, signal B is always sent via channel $Sw1$.

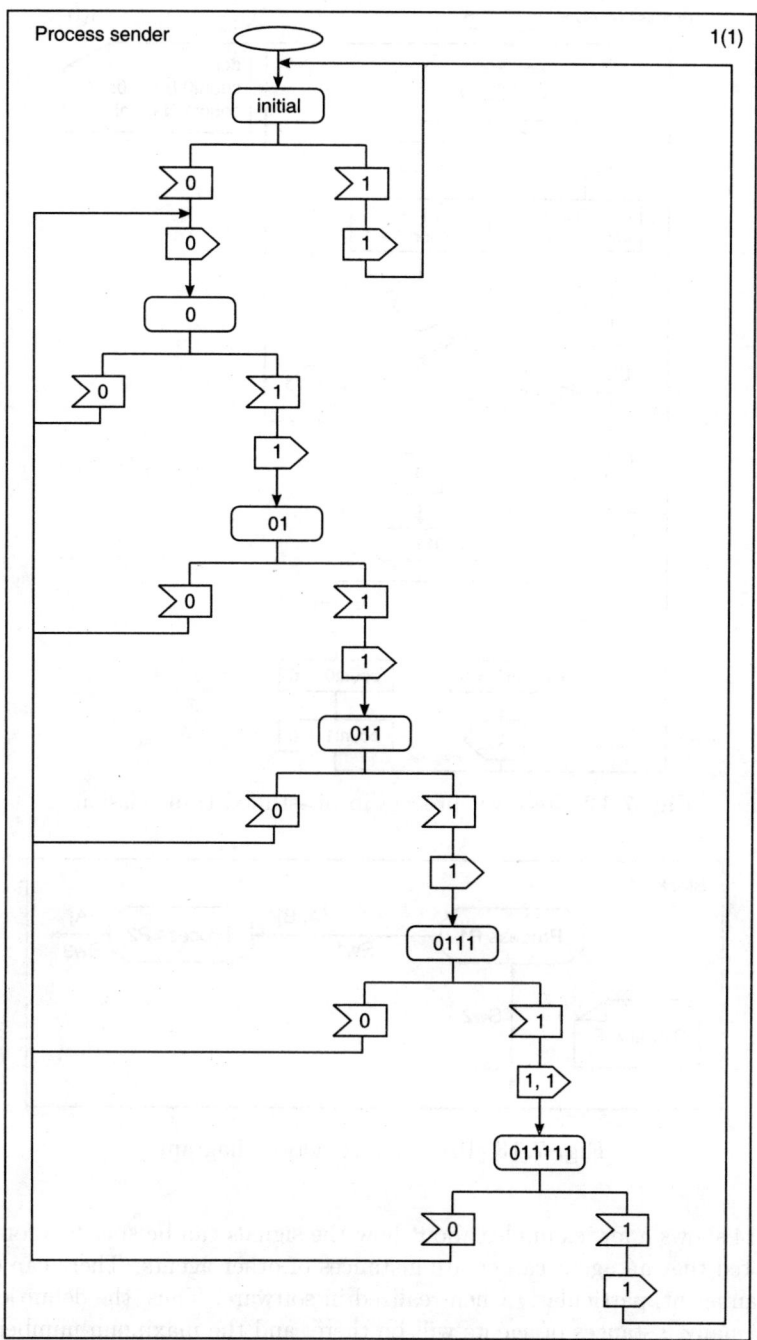

Fig. 7.11 Sender process in bit-stuffed transmission.

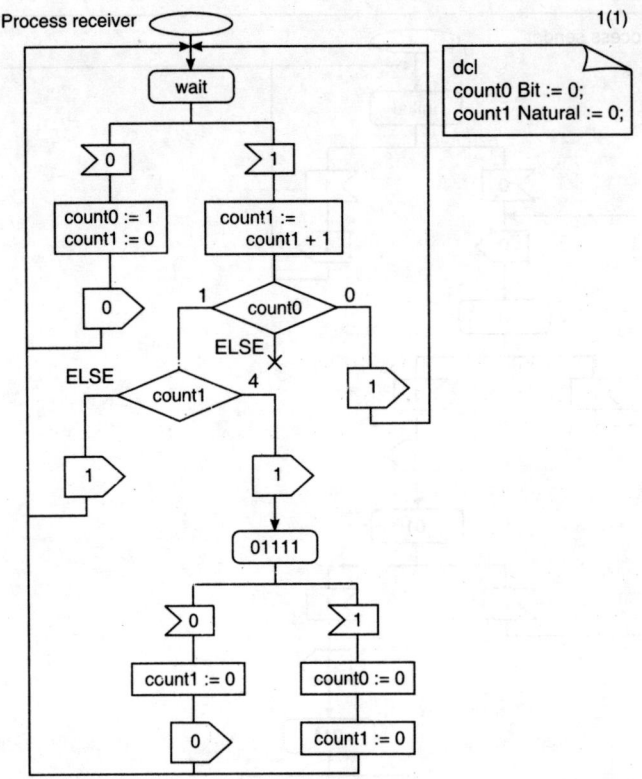

Fig. 7.12 Receiver process in bit-stuffed transmission.

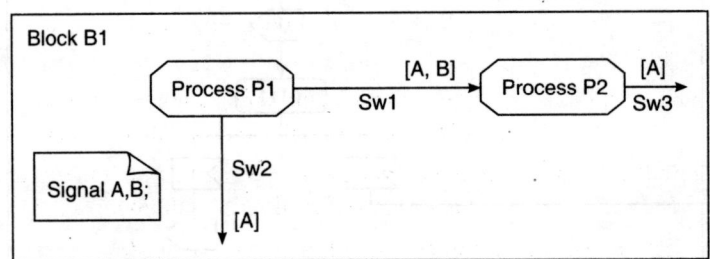

Fig. 7.13 Process interaction diagram.

Figure 7.14 shows a few examples about how the signals can be specified for transmission. It may be noted that an agent can create instances of other agents. There can exist multiple instances of an agent, particularly when realized in software. Thus, the definition of an agent includes how many instances of agents will be there, and the maximum number of instances. The default is one initial instance, and no limit on the maximum. For example, in Fig. 7.15(a), process $P1$ is defined to have 1 initial instance and a maximum of 1 instance, whereas process $P2$ has 1 initial instance and at the most 3 instances. Agents can be created by other agents in a create request (shown in Fig. 7.15(b)) as part of a transition.

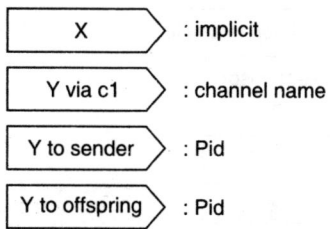

Fig. 7.14 Signal transmission specification.

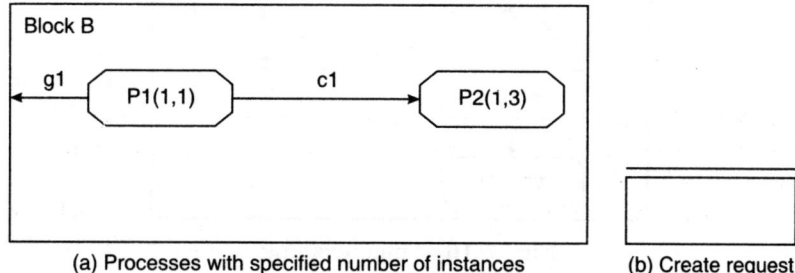

(a) Processes with specified number of instances   (b) Create request

Fig. 7.15 Process creation.

7.3.2 Timer

An agent may have defined *timers* within it. A timer can be created by a definition like,

$$\text{timer } t4 := 10.5;$$

A timer can be started with a *set* and cancelled with a *reset*. For example,

set($\mathbf{now} + 3.2$, $t4$);	– sets the timer $t4$ to 3.2 from current time
set($t4$);	– sets the timer $t4$ to the duration given in the timer definition from current time
reset($t4$);	– resets the timer $t4$ before it expires
active($t4$);	– tests if the timer $t4$ is active

It may be noted that timer definitions are not allowed in state/procedure diagrams. A typical use of timer has been shown in Fig. 7.16.

SDL-2000 contains many other important features, such as procedures, composite states, data types, exception handling, and so on. A detailed discussion on these is beyond the scope of this book. One should refer to the website *http://www.sdl-forum.org* for complete documentation of SDL and find the recent developments.

7.4 Petri Nets

Petri Nets (also called *Place/Transition Net* or *P/T Net*) is one of the several mathematical representations of discrete distributed systems that have been widely used for system modelling in many fields of science. It was invented by *Carl Adam Petri* in 1962 in his Ph.D. thesis.

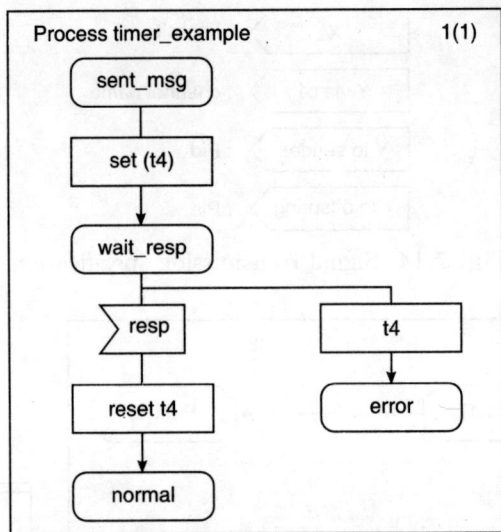

Fig. 7.16 Timer example.

Two important intrinsic features of Petri Nets are *concurrency* and *asynchronous* nature. These, together with generality and flexibility have stimulated their applications in several areas including specification and verification of real-time embedded systems. We will first look into the basic Petri Net structure. However, this cannot be directly used to model embedded systems as it does not have the notion of time. The extensions to this basic structure will be discussed next to show the modelling of embedded systems using Petri Nets.

7.4.1 Basic Petri Nets

A *Petri Net* consists of *places, transitions* and *directed arcs*. Arcs can run between places and transitions. These can never be between places, or betwen transitions. Graphically, the places are represented by circles and transitions by solid lines. The places from which an arc runs to a transition are called *input places* for the transition. Similarly, the places to which arcs run from a transition are called *output places* for the transition. Each place may contain any number of *tokens*. In basic *Petri Net* structure, a token is represented by a small filled circle. The distribution of tokens among the places, at any point of time, is called *marking*. A transition is *enabled* when all its input places have sufficient number of tokens. An enabled transition may *fire*, consuming the tokens from its input places, and placing a specified number of tokens at its output places. The entire firing operation is *atomic*, that is, once a transition fires, all the subtasks, as listed above, are performed together. It may further be noted that the execution of *Petri Nets* is non-deterministic. That is, given a number of enabled transitions at a time, zero or any one of them can fire. It is not necessary for an enabled transition to fire till the infinite time. Due to this non-determinism, *Petri Nets* are suitable for modelling concurrent behaviour of distributed systems. Figure 7.17 shows a very simple *Petri Net* consisting of two places $p1$ and $p2$, and a transition $t1$. Initially, only the place $p1$ has a token. Thus, the transition $t1$ is enabled. Once the transition fires, the token is removed from the input place $p1$ of $t1$ and is put in the output place $p2$.

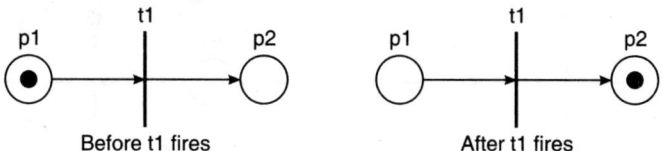

Fig. 7.17 A simple Petri Net.

Formally, a *Petri Net* is a 6-tuple (S, T, F, M_0, W, K), where,

- S is a set of places.
- T is a set of transitions.
- F is a set of arcs with the restriction that it can run between places and transitions only, that is $F \subseteq (S \times T) \cup (T \times S)$.
- $M_0 : S \to N$ (N is the set of natural numbers) is the initial marking, for each place $s \in S$, there are $n \in N$ tokens.
- $W : F \to N^+$ is the set of arc weights. It assigns to each arc $f \in F$ some $n \in N^+$ denoting how many tokens are consumed from an input place by a transition. For the output arcs, the weight indicates the number of tokens to be deposited at the output place.
- $K : S \to N^+$ is a set of capacity restrictions. It assigns a positive integer $n \in N^+$ to each place $s \in S$ denoting the maximum number of tokens that the place can hold.

A weighted *Petri Net* has been shown in Fig. 7.18 before and after the firing of transition $t1$. It may be noted that the weights for edges with no label are taken as unity. Next, we will look into a few examples of system modelling using basic *Petri Nets*.

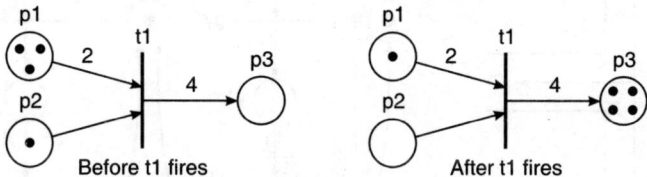

Fig. 7.18 A weighted Petri Net.

Example 7.3

Modelling of a system on an assembly line that counts 5 items and then sends a signal to the operator has been shown in Fig. 7.19. It has four places $p1, p2, p3, p4$ and three transitions $t1, t2, t3$. Their meanings are as follows:

$p1$: item produced on assembly line
$p2$: count items
$p3$: signal is on
$p4$: signal is off
$t1$: sensor recognizes item going by
$t2$: turn signal on
$t3$: turn signal off

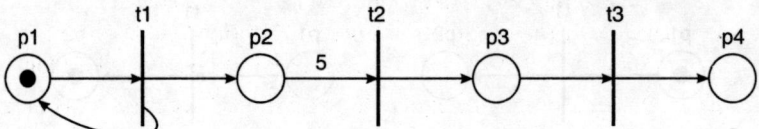

Fig. 7.19 Petri Net of Example 7.3.

Example 7.4

Mutual exclusion: Consider a road-crossing in which two roads (1) and (2) are crossing each other [see Fig. 7.20(a)]. To avoid collision, at any point of time, traffic can be allowed in only one road. Accordingly, signals have been installed, consisting of three lights—*Red (r)*, *Yellow (y)*, and *Green (g)*. To model this system using *Petri Net*, consider the diagram shown in Fig. 7.20(b). The places $r1$, $g1$ and $y1$ represent places corresponding to red, green, and yellow lights being on for road (1). Similarly, $r2$, $g2$, and $y2$ represent the same for road (2). The place x stands for exclusive access permission. The transition $rg1$ can fire, provided the places $r1$ and x have tokens. The rest of the diagram can be followed easily. The places $g1$ and $g2$ cannot get tokens simultaneously, ensuring that traffic is allowed only in road (1), only in road (2), or in none at any point of time.

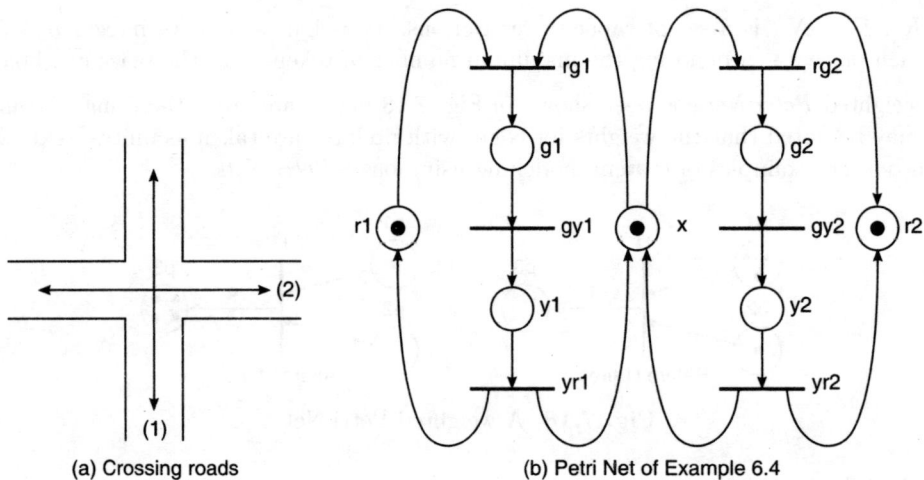

(a) Crossing roads   (b) Petri Net of Example 6.4

Fig. 7.20 Petri Net modelling for mutual exclusion.

Example 7.5

Readers–Writers synchronization: This is another classical problem on synchronization, in which there are k number of tasks accessing a shared resource. Out of these k tasks, some are *readers* that want to read the shared item, but do not want to modify it. The remaining are *writer* tasks that modify the shared item. Naturally, there is no restriction on the number of reader tasks proceeding simultaneously, however, only a single writer can access the item. When a writer task is accessing the shared item, no other reader or writer task can be allowed to access it. The Petri Net corresponding to this has been shown in Fig. 7.21. It may be noted that for a writer to be allowed, it consumes one token from the place *Processes* and k tokens from the *Repository*. On the other hand, for a reader to be allowed, it needs one token each from *Processes* and *Repository*. On finishing, a writer replenishes one token to *Processes* and k to the *Repository*, while a reader replenishes one token to each place.

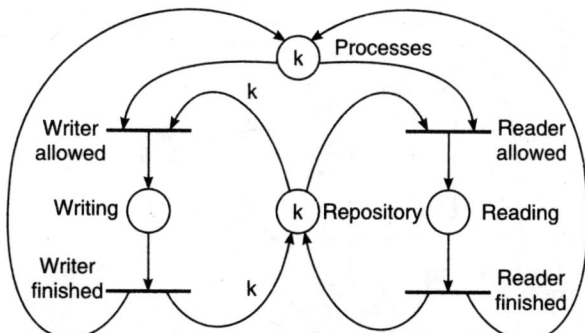

Fig. 7.21 Readers–Writers synchronization of Example 7.5.

Example 7.6

Communication protocol: In this example, there are two processes—*Process1* and *Process2*. *Process1* sends messages to *Process2* and *Process2* acknowledges. After sending a message, *Process1* waits for an acknowledgement from *Process2*. The Petri Net model of the protocol has been shown in Fig. 7.22.

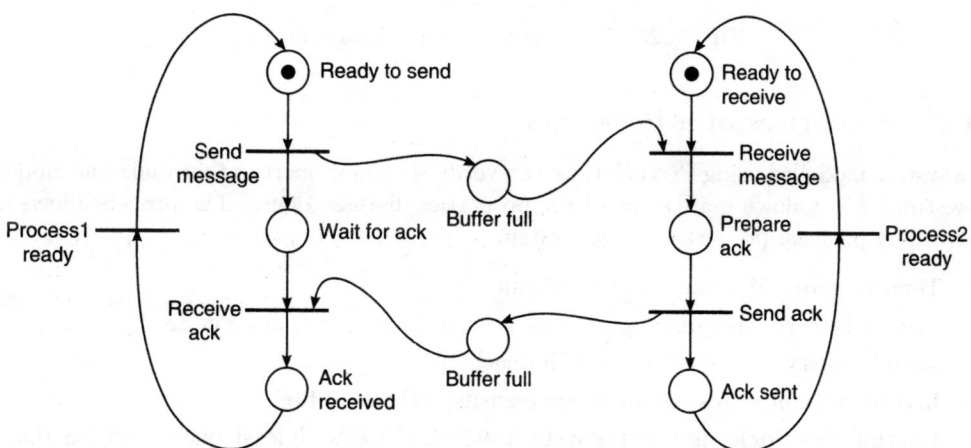

Fig. 7.22 Communication protocol of Example 7.6.

Example 7.7

A vending machine: This example considers a chocolate vending machine that sells two types of chocolates worth Rs. 15 and Rs. 20. It takes as input two types of coins, one worth Rs. 5 and the other worth Rs. 10. There are two more buttons to input the type of chocolate asked for. After entering the requisite amount of money, the user inputs the type and the chocolate is dispensed. The Petri Net model for the system has been shown in Fig. 7.23.

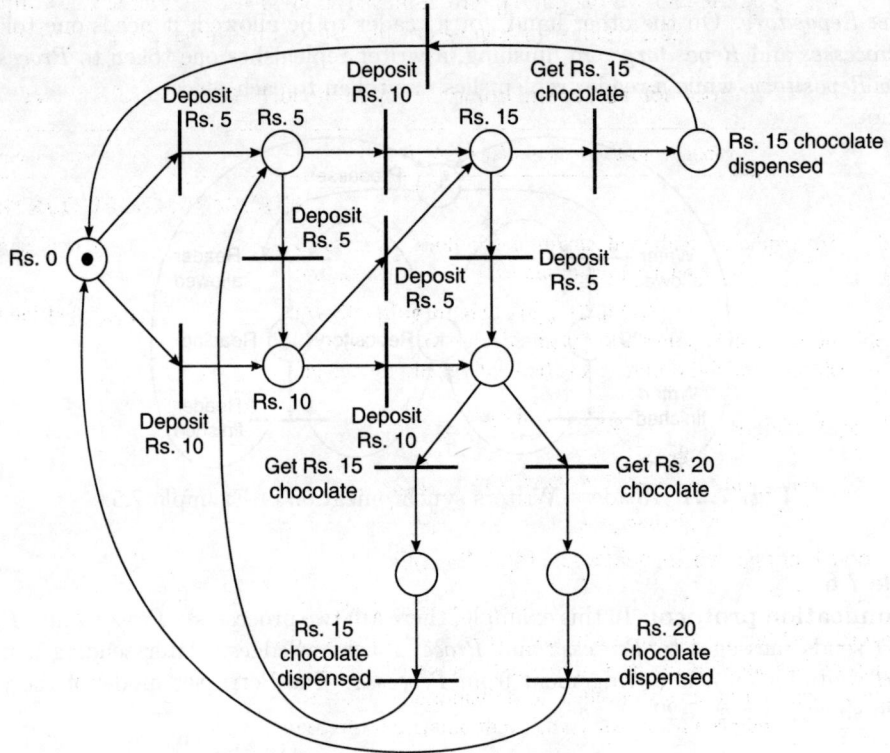

Fig. 7.23 Vending machine of Example 7.7.

7.4.2   Properties of a Petri Net

On a system modelled using Petri Net, we can verify several properties that make the modelling interesting. The following are some of the properties discussed here. The process allows us to verify many physical properties of the system.

1. **Termination**: Whether the net terminates.
2. **Immediate reachability**: Is a state reachable when a transition fires?
3. **Reachability**: Is a state eventually reachable?
4. **Liveness** in all states: Is there one transition that can fire?
5. **Partial deadlock**: Is there a state in which there is at least one transition that can never fire?
6. **Deadlock**: Is there a state in which none of the transitions can fire?
7. **Safety** in all states: Does each place contain at most one token?
8. **Boundedness** in all states: Is there a limit to the number of tokens that can be in one place?
9. **Conservativeness**: Is the total number of tokens in the Petri Net constant?

In the following, we discuss about some of these important properties:

1. *Reachability:* The state of a Petri Net is identified via its *marking*. As noted earlier, marking corresponds to a distribution of tokens amongst the places of the net. A reachability analysis answers the question, whether a given marking (state) is reachable from another specified marking (state). For example, Fig. 7.24 shows two markings M and M' of a net. It can be argued that M' is not reachable from M.

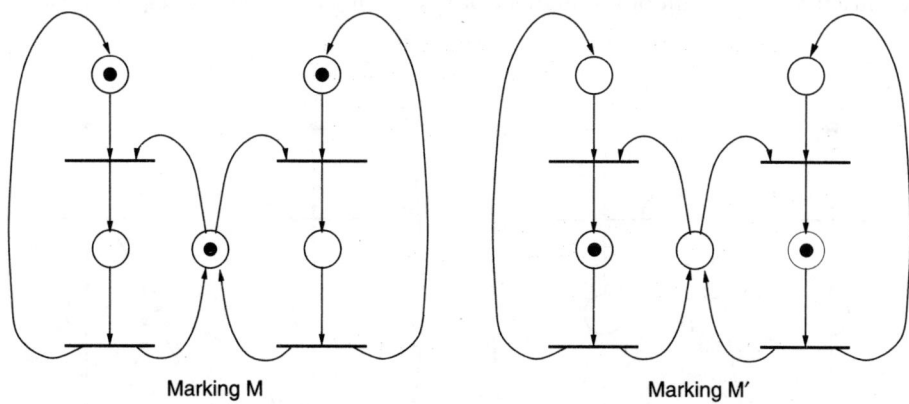

Fig. **7.24** Reachability example.

2. *Liveness:* A transition T is said to be live, if for every marking M_1 reachable from initial marking M there exists a marking M_2 that enables T and is reachable from M_1. The total net is live if all its transitions are live. Figure 7.25(a) shows a net which is not live in the sense that if transition T_1 fires, there is no chance of T_2 to fire in future. However, the net shown in Fig. 7.25(b) is live.

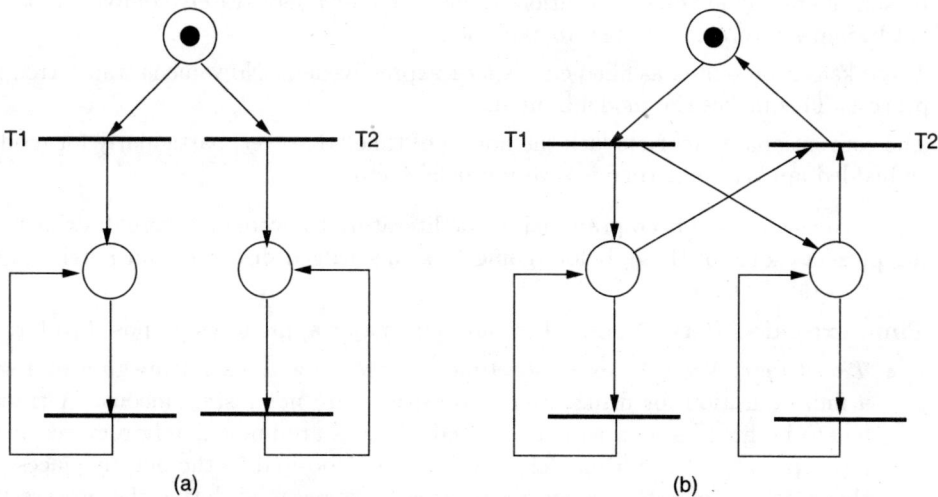

Fig. **7.25** Liveness example.

3. *Boundedness:* It refers to the maximum number of tokens that each place of a net can contain. For the net shown in Fig. 7.26(a), the place P_1 can hold any number of tokens.

Each time the transitions on the left of P_1 fire, one token is deposited into P_1. Thus, the net is not bounded. However, for the net shown in Fig. 7.26(b), the places P_1 and P_2 each can hold at most two tokens, whereas the other places can hold just one token. Thus, the net is bounded. A net is said to be *safe* if its bound is 1. Figure 7.26(b) also shows the property of *place invariants*. The places P_1 and P_2 always satisfy the condition that the sum of the number of tokens in them is always equal to two.

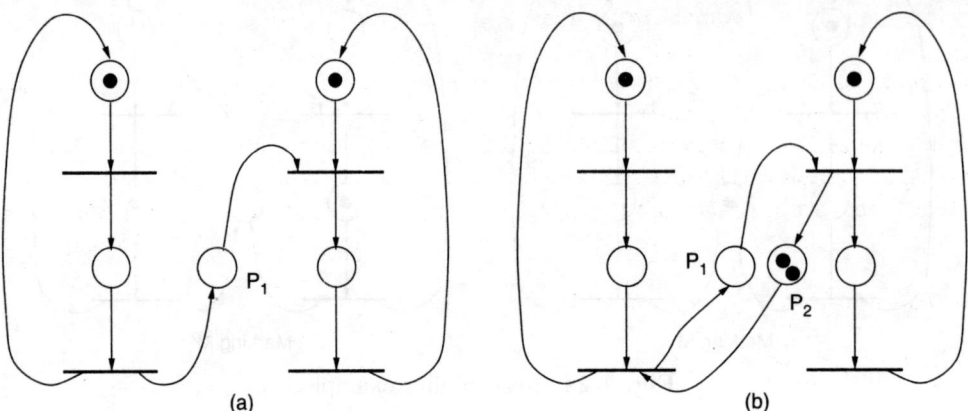

Fig. 7.26 Boundedness and place invariants example.

7.4.3   Extensions to Petri Nets

The following are the major shortcomings of conventional Petri Nets discussed so far:

1. It lacks hierarchy in the specification. Thus, even for a moderately complex system, the net becomes too big, due to state-explosion.

2. The tokens represented as filled circles lack expressiveness. No value is transferred in the process. This makes the model limited.

3. The conventional Petri Nets lack the notion of time. However, particularly for modelling embedded applications, time is a very crucial factor.

Several extensions have been proposed in the literature to overcome these drawbacks. The following presents some of these, before going to a full-fledged discussion on Petri Net based models of embedded systems.

1. **Time extended Petri Nets:** There are four major approaches proposed so far:
   - *Timed Petri Nets:* An execution time is attached with each transition to associate a finite duration for firing. Here, transitions are not instantaneous. A transition has to be fired as soon as it is enabled. Tokens are immediately removed from its input places. After a time delay, tokens are deposited in the output places. Thus the state of the system is not always clearly represented during the process.
   - *Time Petri Nets:* Here, two values of time are associated with each transition starting from the moment the transition is enabled. One is the minimum time value and the other is the maximum time value. Between these two times, the transition has to fire, unless it is disabled by the firing of another transition.

- *Timed Place Transition Nets:* Time information is associated with places, rather than transitions. This means, each place has got some associated delay. A token must remain in that place for this time interval, before being removed by a transition.
- *Tokens holding time information:* A token has got a time-stamp about when it was created. We will discuss this scheme in detail in the following sections.

2. **Coloured Petri Nets:** These were introduced in late 1970s and have a very strong mathematical theory developed around them. Here tokens often represent objects (such as resources, goods, humans, etc.) in the modelled system. The tokens are thus *coloured*, that is, *typed*. The value of the token is called its colour. Transitions use the values of the consumed tokens to determine the values of the produced tokens. Thus, a transition describes the relation between the values of the input tokens and the values of the output tokens. The transitions may also have some preconditions that take colour of token to be consumed into consideration. Thus, coloured Petri Nets can also be used to model time through token values.

7.4.4 Embedded System Modelling with Petri Nets

There are many works reported to model embedded systems using extended Petri Nets. In this direction, we will look into two important techniques to specify embedded systems. These are:

1. *PRES*, Petri Net-based Representation for Embedded Systems
2. Dual transitions Petri Nets

PRES

PRES is an extension to the classical Petri Nets. It captures explicitly the timing information, allows a hierarchical representation of a system. Tokens are allowed to carry information. PRES can also express concurrency and sequential behaviour. A formal description of PRES is as follows:

PRES is a five-tuple (P, T, I, O, M_0), where,

$P = \{p_1, p_2, \ldots, p_m\}$ is a finite non-empty set of places.
$T = \{t_1, t_2, \ldots, t_n\}$ is a finite non-empty set of transitions.
$I \subseteq P \times T$ is a finite non-empty set of input arcs defining the flow relation between places and transitions.
$O \subseteq T \times P$ is a finite non-empty set of output arcs defining the flow relation between transitions and places.
M_0 is the set of initial marking.

The tokens are also modified. A token is defined to be a pair $k = (v_k, r_k)$, where v_k is the token value and r_k is the token time. The token time is a positive real number representing the time-stamp of the token. For any place, the type of token that it can hold is identified by a *type function* $\tau : P \to K$. The *pre-set* of a transition t, denoted as $°t = \{p \in P | (p, t) \in I\}$ is the set of input places of t. Similarly, the *post-set* of a transition t, denoted as $t° = \{p \in P | (t, p) \in O\}$ is the set of output places of t. A transition has got associated with it, the following:

1. *Output functions:* For every place in t° of a transition t, there exists an output function mapping the token values at places of $^\circ t$ to the token value at the output place. That is,

$$\forall p_j \in t^\circ, \exists f_j : \tau(q_1) \times \tau(q_2) \times \ldots \times \tau(q_a) \to \tau(p_j)$$

with $^\circ t = \{q_1, q_2, \ldots, q_a\}$ and $t^\circ = \{p_1, p_2, \ldots, p_b\}$.

2. *Function delays:* For every function associated with a transition, there is a delay.

3. *Guard:* A guard G_t of a transition t is a Boolean condition that must hold true when all its input places hold token, for the transition to be enabled.

A transition is *enabled* if all its preset places are *marked* (that is, contain tokens), output places are empty (excepting those places acting both as input and output), and its guard is asserted. Every enabled transition has got an associated *trigger time* that represents the time instant at which the transition may fire. The trigger time of an enabled transition is the maximum *token time* of the tokens at its preset places. When a transition fires, each of its output places get a token. The *token time* of a token in an output place is equal to the sum of the trigger time of the transition and the delay of the associated output function.

Representing the hierarchy: A transition t and its surrounding places can be substituted by a single Petri Net structure keeping the functionality unaltered. Similarly, a Petri Net structure can be represented by a single transition. This way, a complex structure can be represented in a hierarchical fashion using Petri Net model.

Example 7.8

Figure 7.27(a) shows the PRES representation of the following multiplication algorithm.

```
int mult( int a, int b)
{
        int x, y, z;
        x = a;
        y = b;
        z = 0;
        while (y > 0) {
                z = z + x;
                y = y - 1;
        }
        return z;
}
c = mult(a, b);
```

The model consists of four transitions t_1, t_2, t_3, and t_4. There are six places—A, B, C, X, Y, Z. The arcs are labeled with the function they compute. Thus, the label a from the place A to the transition t_1 indicates that the value of a will come as token value to the transition. All output functions of a transition are assumed to have same delay, marked within the transition. Thus, each of the output functions associated with transition t_3 has a delay of 6 units and so on. Some transitions, like t_3 and t_4 are guarded. Transition T_3 can fire provided its input places X, Y and Z have tokens and $y > 0$. Once it fires, it computes $Z = z + x$ and $Y = y - 1$ with delay 6. When y becomes equal to zero, the guard condition of t_4 is true. It can fire, provided its input places X, Y and Z have tokens. In that case, after a delay of 4 unit, C gets the value of z. It may be noted that t_4 does not need the value of X, however, the place X must have a token at that time for the consistency of operation. The entire multiplication can be represented hierarchically by a single transition as shown in Fig. 7.27(b).

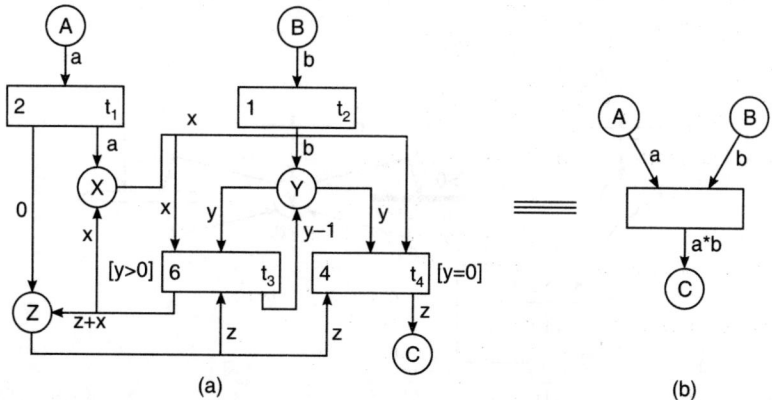

(a)   (b)

Fig. 7.27 PRES structure of multiplication algorithm.

Dual transitions Petri Nets

The concept of dual transitions Petri Nets originated from the fact that an embedded system specification usually consists of both control functions and data operations. The extensions proposed here allow the representation of two mutually exclusive domains—control and data. The structure is a seven-tuple, $N = (P, T, Q, F_C, F_D, W, G)$, where,

$P = \{p_1, p_2, \ldots p_n\}$ is a finite set of places.

$T = \{t_1, t_2, \ldots t_m\}$ is a finite set of control transitions.

$Q = \{q_1, q_2, \ldots q_h\}$ is a finite set of data transitions.

$F_C \subseteq (P \times T) \cup (T \times P) \cup (P \times Q)$ is a set of arcs describing the control flow relation.

$F_D \subseteq (P \times Q) \cup (Q \times P) \cup (P \times T)$ is a set of arcs describing the data flow relation.

$W : \{F_C \cup F_D\} \to Z$ is a weight function, Z being the set of positive and negative integers including zero.

$G : 2^P \to \{0, 1\}$ is the guard function.

The graphical representation of the structure consists of circles representing places, solid bars representing control transitions, rectangles representing data transitions. Arcs represent either control-flow relation (dashed directed arcs) or data flow relation (directed arcs). The concept of *pre-set* and *post-set* remain unaltered. However, as shown in Table 7.1, for control and data transitions, the sets are represented differently.

Table 7.1 Symbols of pre- and post-sets

Domain	pre-set	post-set
Control	$^\bullet x$	$x^\bullet$
Data	$^\circ x$	x°

The multiplication of Example 7.8 has been shown in Fig. 7.28. To perform multiplication, first values are placed at places p_1 and p_2. Thus, data transitions q_1 and q_2 are enabled. Firing them leads to movement of the two values to the places p_3 and p_4. Some operations mapped onto data transitions require a control signal to be activated. For example, transition q_4 is enabled when p_4 holds a token.

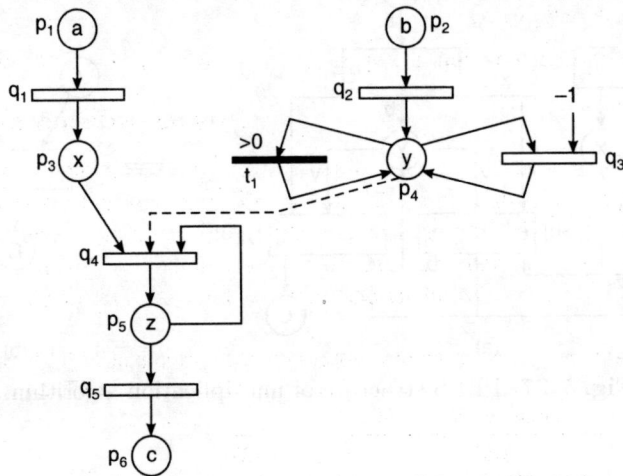

Fig. 7.28 Dual transitions Petri Net structure of multiplier algorithm.

7.5   Unified Modelling Language (UML)

The *Unified Modelling Language* (UML) was originally proposed for modelling complex, software intensive systems. It provides quite a few advantages over *SDL*. It is less formal, and thus can be used at an early stage to structure and analyse the concepts of an application domain before the functional design is made. *SDL* does not support relations in object modelling. Thus, *UML* can be used to develop the formal functional design in *SDL* using the object models of UML.

UML is a graphical language consisting of different types of diagrams that can be used to describe a model from different angles. The diagrams can be classified into three categories:

- *Behaviour diagrams*—describing behavioural features of a system. This includes *activity, state machine, use case*, as well as the following four interaction diagrams.

- *Interaction diagrams*—a subset of behaviour diagrams emphasizing object interactions. This includes *communication, interaction, overview, sequence*, and *timing* diagrams.

- *Structure diagrams*—depict the elements of a specification that are irrespective of time. The constituents are *class, composite structure, component, deployment, object*, and *package* diagrams.

In the following subsections, we briefly describe each of these diagrams.

7.5.1 Activity Diagram

This represents the operational step-by-step workflows of components in a system. It gives the overall flow of control. UML activity diagrams may be considered as object-oriented equivalent of *flow charts* and *data flow diagrams*. Thus, it often consists of different types of components as shown in Fig. 7.29.

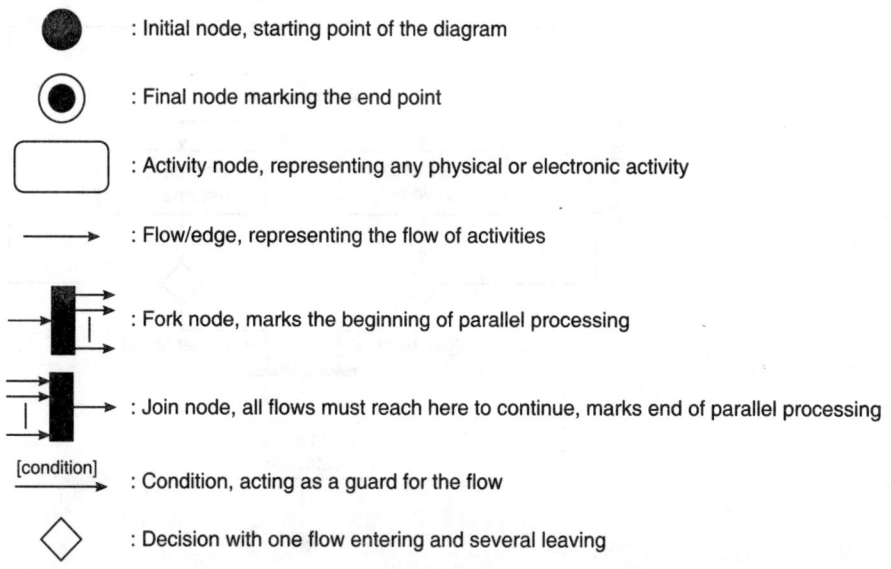

Fig. 7.29 Notations in activity diagrams.

Figure 7.30 shows an example activity diagram for creating a user in a library system. The user fills up a form (hard copy) which is checked for correctness. If found correct, a screen is displayed to the user to enter details. A new user-name (id) and a password are also created.

7.5.2 Class Diagram

This is the main part of object-oriented analysis and design in UML. It can be used to show classes in the system, their interrelations, operations, and attributes. A *class* is shown as a rectangle with three compartments—*name* of the class, *attributes* and *operations*. For example, Fig. 7.31 shows an example class *Door* with a single attribute *status* of type *String*, initialized to *closed*. It has two operations defined on it *Open()* and *Close()*.

An *object* is an instance of a class. A UML object is represented by a rectangle with one or more compartments like class. Figure 7.32 shows an object *D* of class *Door*.

Many kinds of relationships can be modelled in UML. These are as shown in Fig. 7.33.

An *association* defines a relationship between two or more classes resulting in *binary-* or *n*-ary relations respectively. *Aggregation* models the whole/part relationship. A *dependency* is a relationship that indicates that a model element is in some way dependent on another model element. *Generalization* is a relationship between a more general element (superclass) and a more specific element (subclass). It can be used to define *inheritance/hierarchy* as shown in Fig. 7.34.

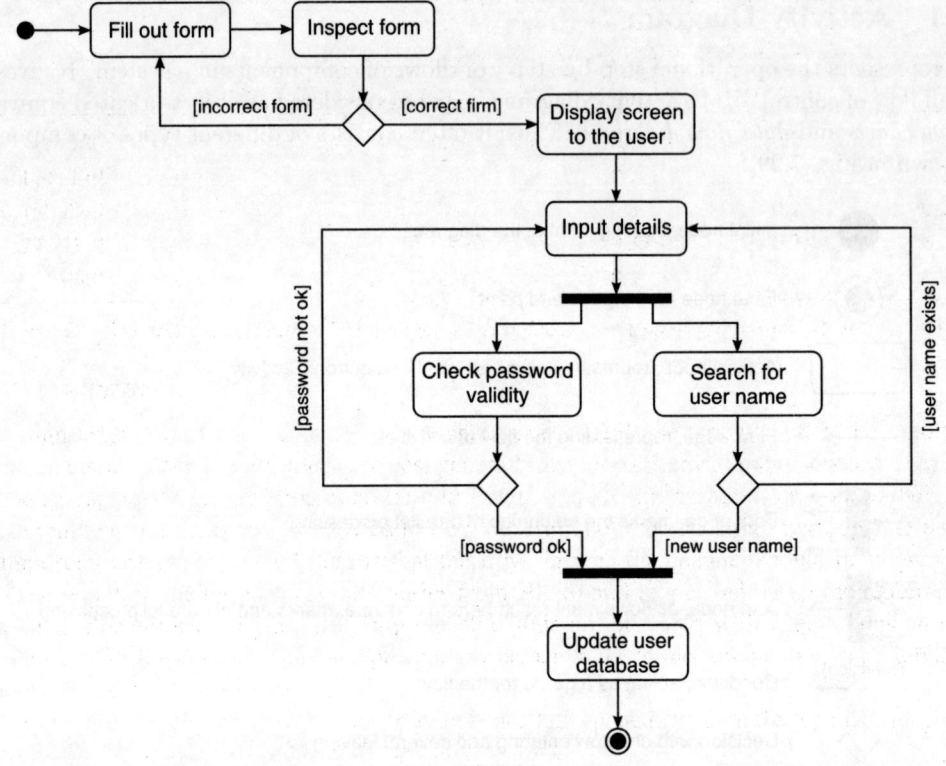

Fig. 7.30 An activity diagram showing user creation in a library system.

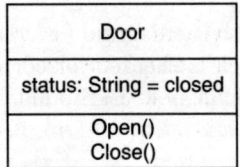

Fig. 7.31 An example class.

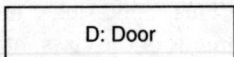

Fig. 7.32 An example object.

7.5.3 Communication/Collaboration Diagram

UML 2 *communication* diagrams (known as *collaboration* diagrams in UML 1.x) are used to show objects and messages involved in accomplishing a purpose or a set of purposes. The message sequence is identified by the sequence numbers. Format of a message is as follows:
precondition/sequence-number [expression] : return-value := message-name(parameter-list)

Figure 7.35 shows a collaboration diagram showing the messages passed between a library user and the various components of the library system.

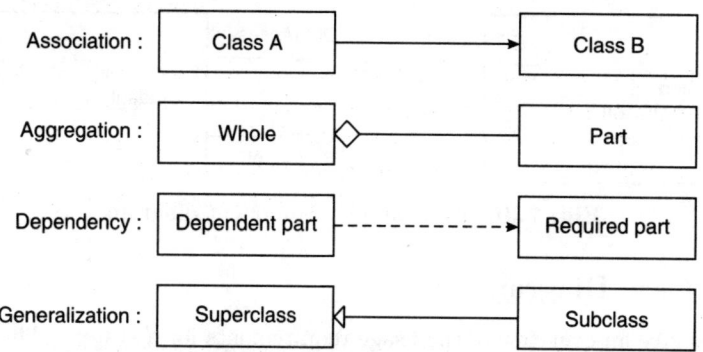

Fig. 7.33 Relationships among UML classes.

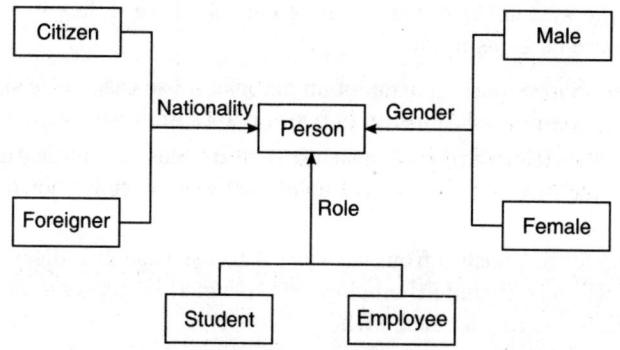

Fig. 7.34 Inheritance among UML classes.

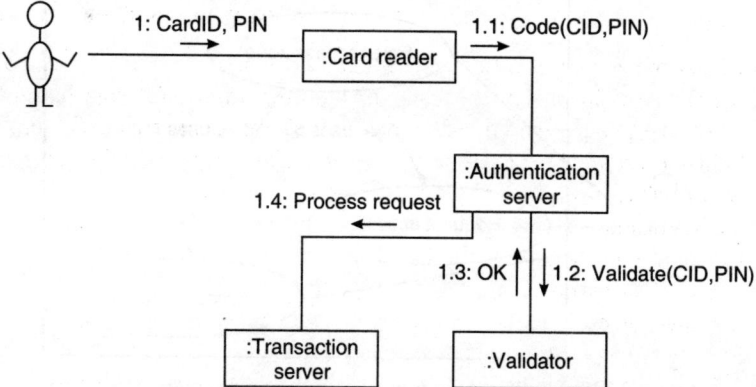

Fig. 7.35 An example communication/collaboration diagram.

7.5.4 Component Diagram

Component diagrams depict how a system is split up into components. They also show the dependency between the components. Figure 7.36 shows a component diagram with ports and interactions.

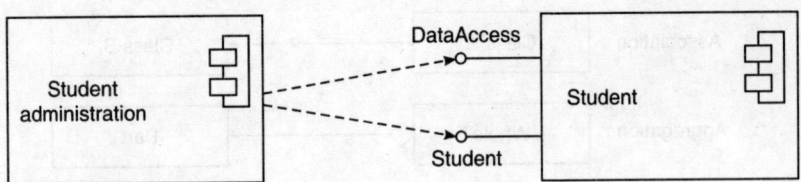

Fig. 7.36 An example component diagram.

7.5.5   Use Case Diagram

These diagrams give an overview of the usage requirements for a system. They show relation-ships among *actors* and *use cases*. An *actor* is a role of an object or objects outside the system that interacts directly with it in a use case. It has class-like properties. A *use case* is a unit of functionality of the system or a class. There can be three different types of associations between actors and use cases as follows:

1. *Communicates:* Shows participation of an actor in a use case. It is shown as a solid line, and is the only possible relationship between actors and use cases.

2. *Extends:* It is a relationship from use case A to use case B, indicating that an instance of use case B may include the behaviour of use case A. It is shown as a generalization arrow labeled with $<< extends >>$.

3. *Uses:* It is also a relationship from use case A to use case B indicating that an instance of use case A will also include the behaviour specified by use case B. This is also shown as a generalization arrow labeled with $<< uses >>$.

Figure 7.37 shows an example of use case diagram.

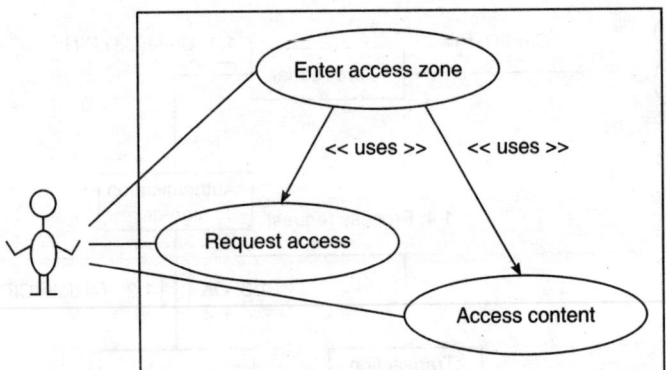

Fig. 7.37 An example use case diagram.

7.5.6   Sequence Diagram

Sequence diagram of a system shows some simple interactions between objects arranged in a time-sequence. For each object, a life-line is shown. Exchange of messages between the objects are also represented. When an object is not active, its lifeline is shown as a dashed line. During active period, the life-line is represented as a thin rectangular bar. An object may become active and inactive several times in its life-span. The life-span of an object may be terminated by putting a × at the end of a life-line. Messages are shown as arrows labeled with name of the message and parameters, if any. Figure 7.38 shows an example sequence diagram.

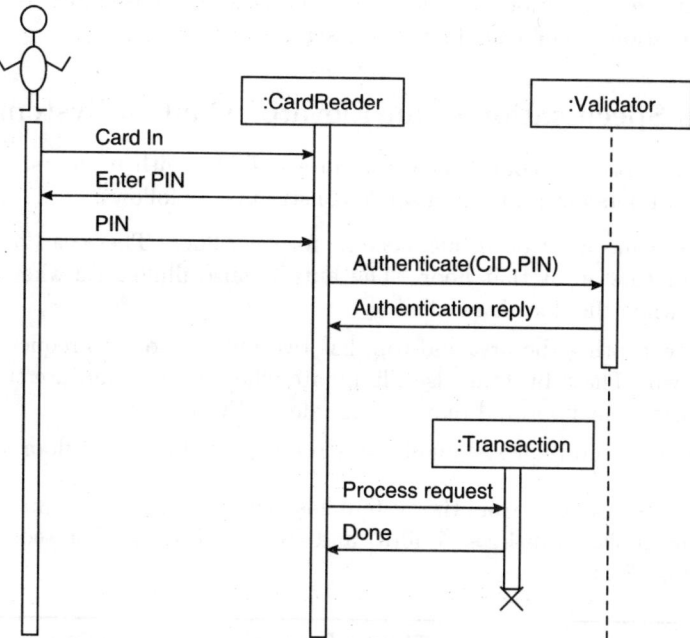

Fig. 7.38 An example sequence diagram.

7.5.7 Other Diagrams

UML contains some more diagrams, which are not very much used in specifications, however, may be utilized to bring clarity to the specification. These are briefly described below:

1. *Composite structure diagram:* This is used to describe the internal structure of a class. It can include internal parts, ports, etc.

2. *Deployment diagram:* The *deployment diagram* of a system models the hardware used in system implementations, components deployed on the hardware and the association between these components.

3. *Interaction overview diagram:* These are variants of UML activity diagrams showing control flow. Nodes within this diagram are *frames*, rather than normal activities. Frames can depict any type of UML interaction diagrams (like sequence diagrams, communication diagrams, timing diagrams, etc.).

4. *Object diagram:* An object is an instance of a class. Object diagrams are similar to the class diagrams excepting the fact that these can be used to explain complex relationships between classes. Objects belonging to the same class may now be associated with objects of other classes in different ways.

5. *Package diagram:* Packages enable us to organize elements into groups. These can be used in conjunction with any of the UML diagrams just like file folders.

6. *State machine diagram:* This is similar to *StateCharts* discussed earlier. This can be used to capture state-based behaviour of a system.

7. *Timing diagram:* These can be used to explore the behaviour of one or more objects throughout a given period of time. The time may be represented as various phases of system operation in embedded system design.

7.5.8   UML Specification of an Elevator Control System

Let us consider a system to control elevators in a building with m floors. We need to move elevators between floors according to a set of constraints, as follows:

1. Each elevator has a set of m buttons, one for each floor. These can be pressed to cause the elevator to stop at that floor. The buttons also illuminate when pressed, and are turned off when the floor is reached.

2. Each floor excepting the first and top, has two buttons, one to request up, the other to request down. These buttons also illuminate when pressed and are turned off when an elevator visits that floor and moves in the desired direction.

3. In case there are no requests, an elevator remains at its current floor with door closed.

The detailed class diagram from the system has been shown in Fig. 7.39. The collaboration diagrams have been shown in Figs. 7.40(a) and (b) respectively. The sequence diagram has been shown in Fig. 7.41.

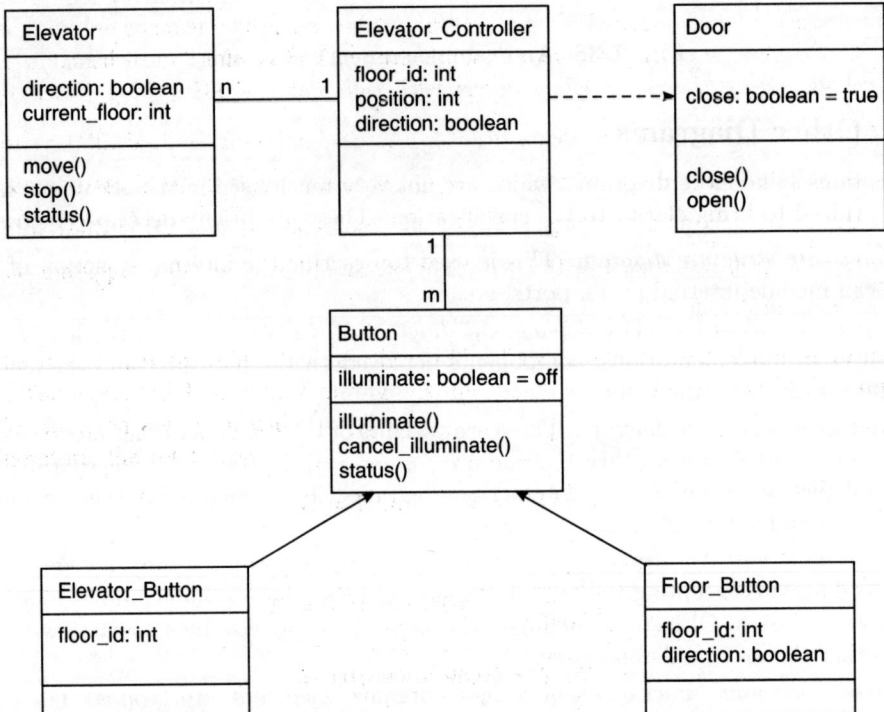

Fig. 7.39 Class diagram.

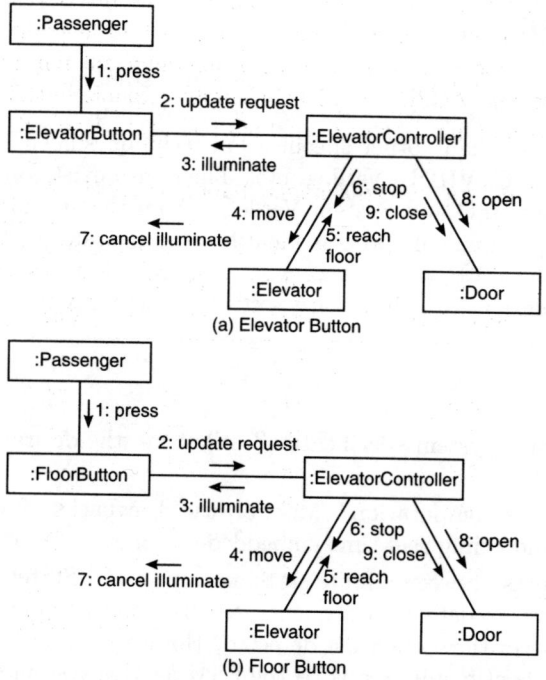

(a) Elevator Button

(b) Floor Button

Fig. 7.40 Collaboration diagrams for serving.

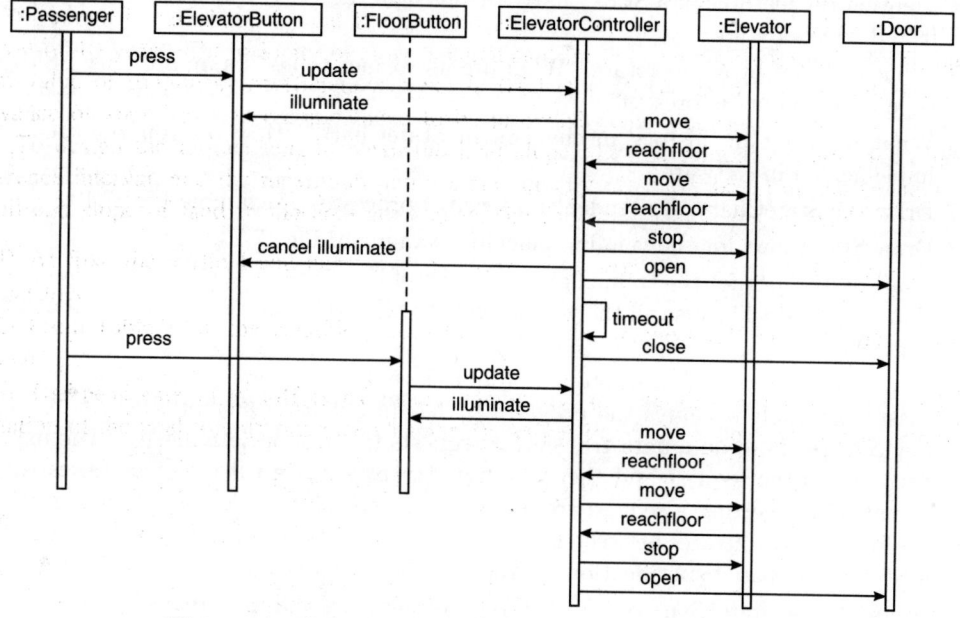

Fig. 7.41 Sequence diagram for elevator operations.

7.6   Conclusion

Specification techniques help in expressing the desired system behaviour. In this chapter, we have seen many such techniques—some are very formal, while others are more pictorial. It may be noted that there exists many other specification techniques such as hardware description languages like Hardware C, VHDL, Verilog, etc. There also exist some system-level description languages, like, SystemC and System Verilog. From the specification of a system, the refinement (automated or manual) process identifies the modules to be realized in hardware or in software. In the next two chapters, we will see how to decide upon this partitioning and verify the partitioned design using hardware–software cosimulation.

Exercises

7.1 What is meant by a system specification? What are the desirable features of a specification technique?

7.2 What is a model of computation? Mention the drawbacks of Von Neumann model of computation in modelling real-time embedded systems.

7.3 State the differences between a finite state machine and a StateChart. Suppose, we have a sequence detector to detect the sequence "100011" in the input bit stream. The system outputs '1' whenever the sequence is detected. However, if another input a is equal to '1', it toggles output continually. Draw (i) the FSM for this system and (ii) the StateChart. From this explain how StateChart is superior to FSM.

7.4 Mention the features of a StateChart. In this light, describe which are advantageous for it and which are not.

7.5 What is meant by a superstate? What are its various types? In an ordinary FSM, what type of superstate is present?

7.6 What is meant by history mechanism in StateChart? How do you think can it be implemented in software?

7.7 Draw the StateCharts for sender and receiver processes of example Fig. 7.10.

7.8 Draw StateChart for the vending machine example of Fig. 7.23.

7.9 Draw StateChart for the traffic light controller of Fig. 7.20. Incorporate a *manual* button that if pressed will cause all the yellow lights to blink, indicating a manual mode of traffic control.

7.10 A system consists of two masters and four slave devices, connected over shared bus. A master normally executes the program by fetching it from memory. Whenever a slave device is ready to do a data transfer, it requests the appropriate master. The master in turn grants the request and the data transfer takes place till the slave raises the *over* signal. Upon getting a request from a device, a master first sends a request to the bus arbiter which arbitrates between the requests and grants the master to use the bus. Draw a StateChart based specification for the overall system.

7.11 Augment the StateChart of Fig. 7.8 for entering hour and minute also.

7.12 Draw the SDL specification for Exercise 7.8.

7.13 Draw the SDL specification for Exercise 7.9.

7.14 Draw the SDL specification for Exercise 7.10.

7.15 Draw the SDL specification for the digital watch of Fig. 7.7.

7.16 Draw the SDL specification for the date entry mechanism of Fig. 7.8 and also for Exercise 7.11.

7.17 State the advantages of SDL over StateChart.

7.18 What is a Petri Net? What is meant by the statement that the firing operation is atomic?

7.19 A producer–consumer problem consists of two processes—a producer and a consumer. The producer produces data items to be consumed by the consumer process. A finite sized buffer in between them acts as an interface between them. The producer should wait once the buffer becomes full, and the consumer has to wait when the buffer is empty. Show a Petri Net based model for the same.

7.20 What are the properties of a system that can be verified from the Petri Net model of a system?

7.21 Draw the Petri Net of digital watch of Fig. 7.7.

7.22 Draw the Petri Net of the date entry mechanism in Fig. 7.8.

7.23 Draw the Petri Net for the modified traffic light controller in Exercise 7.9.

7.24 What are the drawbacks of general Petri Nets?

7.25 Discuss about the different types of time extended Petri Nets.

7.26 What is coloured Petri Net? How does it model time and data types?

7.27 How is an embedded system modelled using PRES? How the same is done in dual transition Petri Nets?

7.28 Write a pseudo-code for computing the greatest common divisor (GCD) of two numbers. Show the corresponding PRES and dual transition net models.

7.29 What are the different types of commonly used diagrams in UML?

7.30 For all the exercises related to modelling of example systems noted earlier, draw the UML diagrams.

CHAPTER 8

Hardware–Software Cosimulation

Simulation is one of the most effective measures to ensure system correctness. Since an embedded system design consists of interacting software and hardware components, it is very much essential that a simulation of the system includes simulating both, along with their interactions. In the past, cosimulation used to be done late in the process, after the hardware has been deemed to be mostly working, i.e., stable, such that it can interact properly with the software. Software developers used to develop their code with limited facility to test, in the absence of the hardware platform. Painful integration efforts were needed towards late in the design cycle, and minor miscommunication became major design flaws. Most of these problems would be patched up in software at the cost of performance or even expensive redesign of hardware/software. Cosimulation at an early stage attempts to solve this problem.

The typical usage of cosimulation during codesign are as follows (Fig. 8.1):

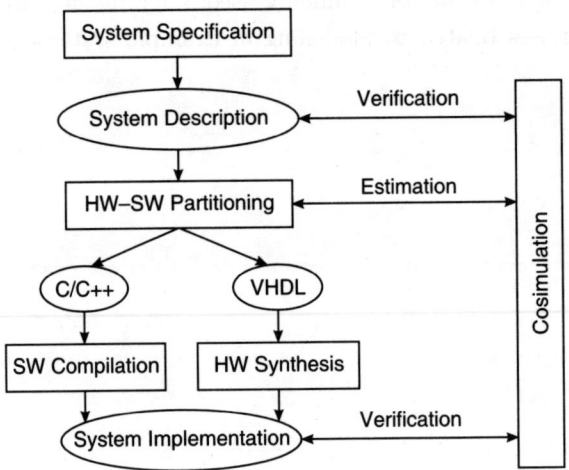

Fig. 8.1 Codesign flow with cosimulation.

- Verification of system specification to check whether the specification is as per the intent of the system designer.
- Verification of system implementation.
- Performance estimation to be used during system partitioning.

It may be noted that cosimulation is used in different subdomains within the domain of electronic system design. Apart from hardware–software cosimulation, it can refer to any other simulation, such as mixed analog/digital systems. All cosimulations perform a simultaneous simulation of two or more parts of a system defined at different levels of abstraction. Some part of the system may correspond to the abstraction of hardware components while some other part may correspond to the software components. The challenge of cosimulation is to design the bridge between different simulators used for different components. A cosimulation tool has to solve the problems of synchronization between the simulators and associated type conversion between the exchanged signals.

8.1 Dimensions in Cosimulation

Cosimulation is the process to simulate different system components together. However, the major problem occurs due to speed mismatch between the real components, which may not be captured in the individual component level simulators. The component level simulators need to interact with each other to exchange signal values. This gives rise to a number of major issues to be resolved in designing a cosimulation environment.

1. *Communication:* Communication between the hardware- and the software-simulators is the most important issue in cosimulation. The communication mechanism between the simulators often determine the overall efficiency (in terms of accuracy and speed of simulation) of cosimulation. Most of the modern operating systems provide facilities for interprocess communication (commonly known as *Interprocess Communication Primitives* or IPCs). The software part, often developed in C-like language supports direct call to the IPC routines. Hardware description languages, like VHDL and Verilog, also support mechanisms to call procedures written in some other languages. VHDL supports the link of procedures, written in C code, through its *foreign language kernel*. The language Verilog supports invocation of C-functions by a *programming language interface* (PLI).

2. *Synchronization:* The control of timing in interactions between hardware and software parts of a system is essential to ensure the correctness of simulation. One possibility in performing the communication and synchronize between the components is to use shared memory. The components read to/write from the shared memory. A read-modify-write mechanism is commonly used to ensure this synchronization. The accesses are totally ordered and the memory location is locked during read/write operation. However, this type of approach results in a tight coupling between the hardware and software components. Their design often becomes dependent on each other. A better mechanism is to use the IPC channel discussed earlier. To control synchronization in such an environment, often a handshake mechanism is used. The mechanism has been shown in Fig. 8.2 in which the C-program assumes the role of the master, initiating a transfer request from the hardware described in Verilog. The master waits for the slave to put the data and send an acknowledgement to the master. Master then revokes the request, followed by the slave revoking the acknowledgement. The PLI part performs the interfacing task, the C program actually talks to this PLI only.

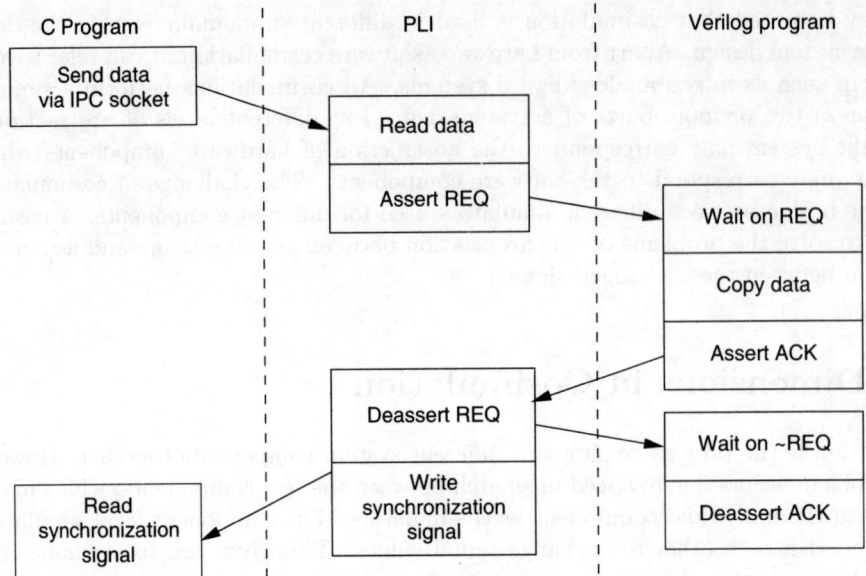

Fig. 8.2 Handshake mechanism between hardware and software.

3. *Scheduling:* Scheduling refers to the synchronization to be achieved between the components—not only their interaction. This is required to ensure proper timing of the actions carried out by the simulators. A concept of *global timing* has to be introduced. Individual simulators may work with *local timing*. Local time is compared with the global time by every simulator. The component with local time equal to the global time should be simulated. Another possibility to achieve this synchronization is to have time stamps for the events. Every event on a signal consists of a value and a time stamp, indicating the time at which the event occurs. A component compares its local time and time stamp to decide which event to consider next.

4. *Models of computation:* As observed in the chapter on *Specification Techniques*, each component has a corresponding model of computation describing its behaviour at certain level of detail. Availability of proper model and its abstraction level controls the accuracy and execution time of the simulator.

8.2   Cosimulation Approaches

There can be several approaches to hardware—software cosimulation ranging from very detailed (for example, gate-level model) to very abstract (such as, instruction-level model). While the hardware simulation can be performed using *Hardware Description Language* (HDL) simulators, software is typically simulated using *Instruction Set Simulator* (ISS). A very simplistic approach for cosimulation could be as follows:

- Using an HDL model of the microprocessor that will run the software part.
- Using HDL models of specific hardware synthesized for the hardware part.
- Integrating all models to perform a full simulation.

This approach is known as *homogeneous simulation* as all the modules are simulated using the same simulator.

A more generic model is the *heterogeneous cosimulation* working on the following principle:

- Use an instruction-set simulator model of microprocessor that runs the software.
- Use HDL model for simulating the hardware.
- Create communication between the simulators.
- Simulators run separately excepting when transferring data.

In some sense, a heterogeneous simulator networks different types of simulators as shown in Fig. 8.3. Heterogeneous simulators incur high overhead due to the voluminous data transfer between hardware and software simulators. The percentage of time that the software accesses the hardware is called the *hardware density* of simulation. Heterogeneous simulation affords better ability to match a task with a tool. A typical example of such a simulator is *Synopsis's Eaglei* that lets the hardware run in many simulators and the software on the native PC/workstation or in instruction-set simulator (ISS). For the hardware portion, HDL simulators are generally used. However, simulation runs at a much slower rate. Thus, it may be difficult to catch many of the practical problems, particularly, the timing related problems. For such a situation, in the absence of real hardware, a prototype realized on FPGA may be used. Such a concept is known as *emulation*. It is a special simulation environment with hardware. The following are some of its features:

- An FPGA implementation can run the whole design as compared to HDL simulators where only a part may be simulated. HDL simulation may be very much time consuming.
- Emulation is expensive compared to pure software simulation, as FPGA chip and associated board are necessary.
- It may run at 10% of real-time, thus much closer to the actual hardware speed. The software simulation runs rather slowly.
- It allows designers of large products to find problems that cannot be found in simulator, as the simulation of full hardware may not be carried out due to time limitations.
- Real devices can be attached with the FPGA based hardware emulators. This helps in visualizing entire system and check its functional correctness.

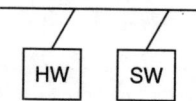

Fig. 8.3 Heterogeneous simulation.

8.3 A Typical Cosimulation Environment

There are several cosimulation techniques reported in the literature. They work on the principle of interprocess communication between two or more heterogeneous simulator processes. The cosimulation environment discussed here consists of three major components—a software process running *C*-programs (software part), a simulation process executing the hardware model in VHDL (hardware part), and an interface model of the communication between software and hardware parts. An optional custom board may be used to accelerate the simulation process by emulating the hardware part on FPGA. The structure has been shown in Fig. 8.4.

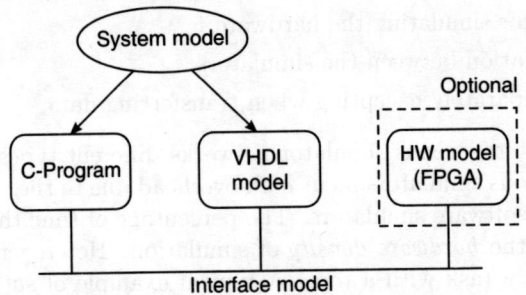

Fig. 8.4 Cosimulation environment.

The cosimulation can be classified into two categories depending upon the time at which it is invoked—before target selection or after target selection. The cosimulation performed before the target has been selected is called *abstract-level cosimulation*, whereas the other one is known as *detailed-level cosimulation*.

8.3.1　Abstract-level Cosimulation

At this level, only the C-program for the software part and VHDL description for the hardware part are available. To communicate between these two corresponding simulators, *interprocess communication primitives* (IPCs) are used. It is assumed that calls to IPC routines have been inserted at appropriate places in both the C-program and VHDL code. Thus, the IPC routines provide the communication channel, i.e., interface, in the absence of any concrete implementation of either software or hardware module. Figure 8.5 shows the structure of cosimulation at the abstract level. All IPC routine calls in a hardware component are grouped into an IPC handler process in VHDL, whereas for a software component, the calls are distributed in the C-program.

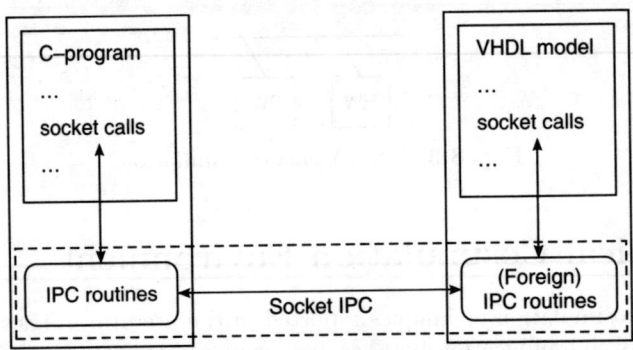

Fig. 8.5 Cosimulation environment at abstract level.

8.3.2 Detailed-level Cosimulation

The detailed cosimulation is used after the system designer has selected a target architecture for both software and hardware. The interface protocol has also been selected. The implementation of interface includes the automatic generation of *decoder/signal register* and insertion of *channel unit* model from library. The structure has been shown in Fig. 8.6.

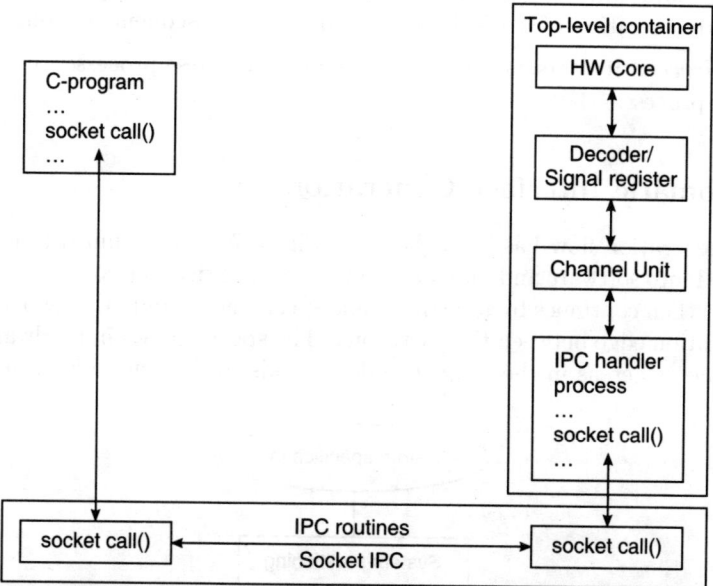

Fig. 8.6 Cosimulation environment at detailed level.

The overall cosimulation process proceeds by the software part (the *C*-program) sending relevant data and *start* signal to the VHDL simulation process. The *C*-program stops and awaits *finish* signal from the VHDL simulation process. On receiving this signal, again the software part continues.

8.3.3 Interface Issues

It is very important that the cosimulation environment provides a transparent communication interface between the software and the hardware modules regardless of target architectures and communication protocols. Hardware module, realized as a VHDL model, should not have any IPC calls within it, it should still visualize the signals coming to/from software as input/output ports only. Moreover, it should be possible to synthesize the interface elements, given the software and hardware models. As shown in Fig. 8.6, the interface elements are— IPC handler, Channel unit, Decoder/Signal register, Top-level container. Their roles are as follows:

1. The *IPC handler* is responsible for reading/writing data from/to the software process using *foreign* IPC routines. These routines are called *foreign* as they are not VHDL code. The VHDL simulator contains a mechanism to invoke such foreign routines. The

IPC handler handles IPC jobs via handshaking. This is often a manually crafted VHDL process that is created during abstract-level simulation.

2. *Channel unit* represents the abstract simulation model of a physical channel device (for example, a DMA controller). It is often selected from an interface library.

3. The *Decoder/Signal register* is inserted between the hardware core and channel unit to match the data transfer width, overcome pin limitation and protocol differences. An automated generation of it will be discussed in the subsequent section.

4. The *Top-level* entity acts as a *Container* holding all these processes along with the core hardware process.

8.3.4 Automatic Interface Generation

The interface generation flow has been shown in Fig. 8.7. It is assumed that the system has been partitioned into software and hardware partitions and that the corresponding flow-graphs are available. It then continues by inserting some special nodes in both the parts—one pair for each communication edge between the partitions. The special nodes in hardware are converted to *signal registers*, whereas in the software side, it leads to the synthesis of *driver functions*.

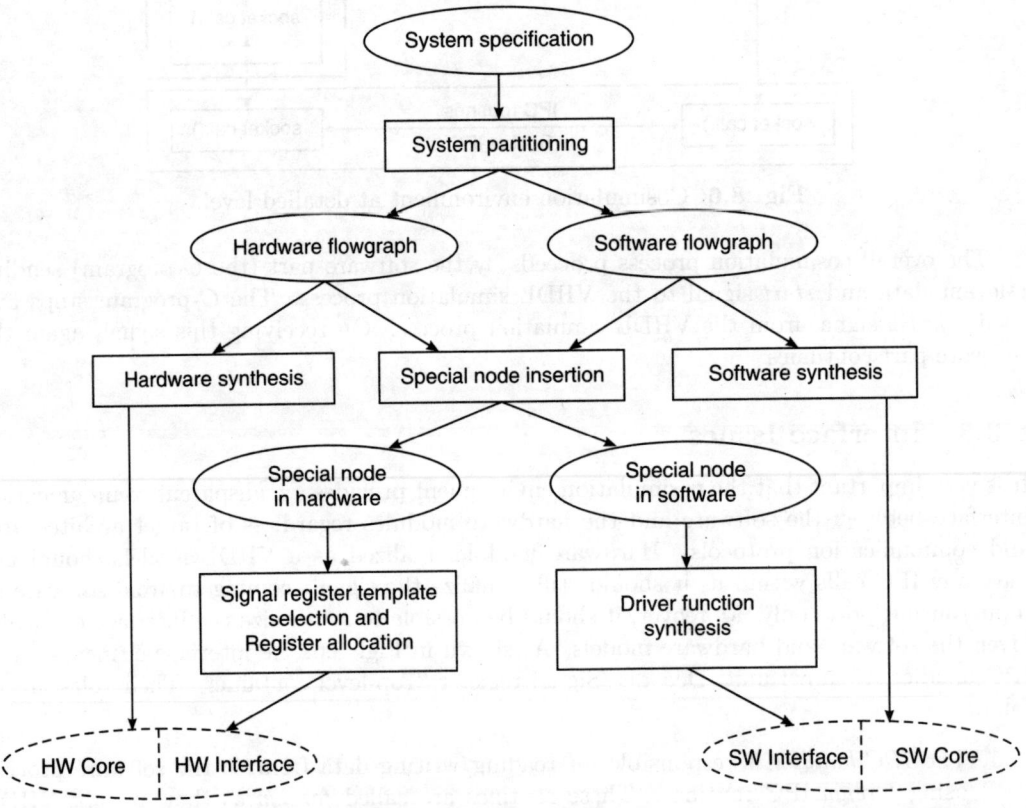

Fig. 8.7 Interface generation.

An example of generated interface has been shown in Fig. 8.8. It is assumed that the core hardware has got three 32-bit input ports a, b, c, and a 32-bit output port d to interface with the software. In the interface module, there is a *protocol converter* that interfaces with the system bus which may be performing data transfer using some different protocol. However, the core hardware should see all the signals as I/O ports only. The protocol converter generates *read/write* signals for the *signal register*. The signal register consists of a set of registers for the input ports of the hardware core, address decoder to enable these registers. For the output ports, it contains output buffers and multiplexers. It is assumed that the system bus is 16-bit wide, thus there are two 16-bit registers for each 32-bit input port of the hardware. The output d is also phased over two 16-bit transmissions.

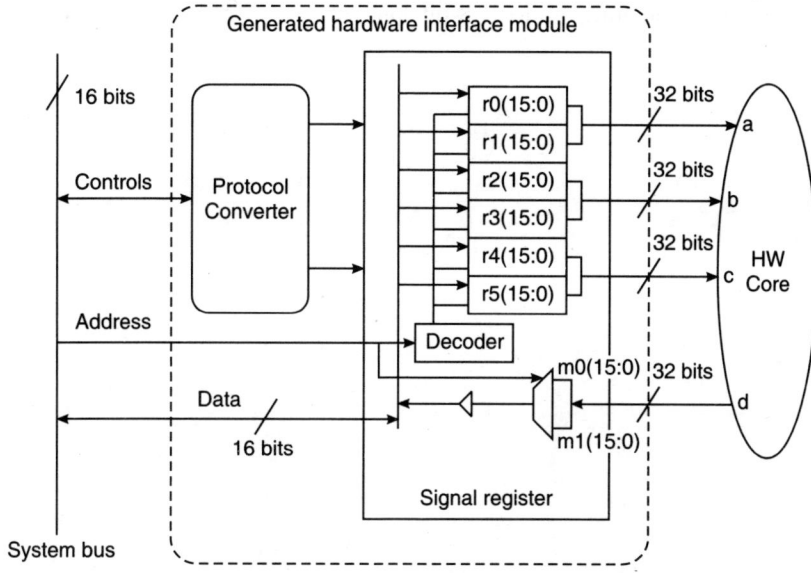

Fig. 8.8 A generated hardware interface.

8.4 Conclusion

In this chapter, we have seen various techniques for hardware–software cosimulation. Heterogeneous cosimulation, preferably augmented by emulation hardware is desirable over a homogeneous simulation. It also necessitates modelling of the interface module in both hardware and software. In the next chapter, we will see how to partition a given system specification into hardware and software for the purpose of cosynthesis.

Exercises

8.1 What is cosimulation and how is it different from simulation of hardware and software systems?

8.2 What are the requirements of cosimulation in the codesign process?

8.3 State the difference between a homogeneous and a heterogeneous cosimulation strategy.

8.4 Explain the role of PLI in cosimulation.

8.5 What is emulation? Mention the features of emulation.

8.6 How does a typical cosimulation environment look like?

8.7 State the basic principle of abstract level cosimulation.

8.8 How does a detailed cosimulation approach differ from an abstract cosimulation approach?

8.9 Discuss the strategy for automated interface synthesis.

8.10 How will the modified interface look like if the individual registers within the signal register are (i) 32-bit long, and (ii) 8-bit long?

CHAPTER 9

Hardware–Software Partitioning

Embedded systems typically consist of application-specific hardware part (containing ASIC, FPGA, etc.) and software part running on general purpose processors, digital signal processors (DSPs), etc. Advancement on both the fronts—high level synthesis to produce custom hardware, and software development and compilation techniques to exploit the features of general processors to their limits, has ushered in a new challenge in embedded system design. Thus, a compiler for embedded system cannot only be a silicon compiler (converting a specification to hardware) or a highly optimized language compiler for a target processor, it should also be able to partition the specification into a number of software and hardware modules. Such partitioners must be extremely fast as they have to explore a good amount of design space in terms of hardware, software, and communications between them. The criteria to optimize may be implementation area, performance, power consumption, etc., or a mix of a number of such factors.

There are quite a few approaches to solve this partitioning problem. Some of these are as follows:

1. Integer Linear Programming (ILP)
2. Heuristic approaches, such as, Kernighan-Lin algorithm
3. Meta search techniques, like
   (a) Genetic algorithm (GA)
   (b) Simulated annealing (SA)
   (c) Particle swarm optimization (PSO)

9.1 Partitioning Using Integer Programming

Integer programming is a mathematical tool to solve constrained optimization problems. It has been used by many researchers to solve the hardware–software partitioning problem as well. To start with, let us consider a constrained optimization example noted below:

Minimize the expression $C = 7x_1 + 3x_2 - 4x_3 + 2x_4$ subject the constratints that each of the x_i's can pick up a value of 0 or 1 and that their sum must be less than or equal to 3. The integer programming formulation is given below:

$$Minimize \quad C = 7x_1 + 3x_2 - 4x_3 + 2x_4$$

such that,

$$x_1 + x_2 + x_3 + x_4 \geq 3$$
$$0 \leq x_1 \leq 1$$
$$0 \leq x_2 \leq 1$$
$$0 \leq x_3 \leq 1$$
$$0 \leq x_4 \leq 1$$

In this case, the solution is $C = 1$ with variable assignment $x_1 = 0, x_2 = 1, x_3 = 1, x_4 = 1$.

In the following a formulation of the hardware/software partitioning problem using integer programming has been presented. For this, the following definitions are needed:

Definition 9.1

*The **target architecture** consists of an ASIC h, a set of processors $P = \{p_1, \ldots, p_{n_P}\}$, external memory and buses between them. The set of target architecture components is defined as,*

$$TA = \{h\} \cup P$$

The ASIC may be considered as the first element of TA with index 0, followed by the processors:

$$ta_0 = h; ta_k = p_k, \forall k \in \{1, \ldots, n_P\}$$

Definition 9.2

*A **system** is defined as a tuple, $S = (EN, V, E, I)$ with the following definitions:*
  $EN = \{en_1, \ldots, en_{n_{EN}}\}$, the set of entities,
  $V = \{v_1, \ldots, v_{nv}\}$, set of nodes, representing instances of entities,
  $E \subset V \times V$, set of edges representing interconnections between instances,
  $I : V \to EN$, $I(v_j) = en_l$, defines that v_j is an instance of en_l.

The following cost metrics are defined for each entity en_l.

1. $c^a(en_l)$ represents the hardware area.
2. $c^{th}(en_l)$ represents the hardware execution time.
3. $c^{dm}(en_l)$ represents the used software data memory.
4. $c^{pm}(en_l)$ represents the used software program memory.
5. $c^{ts}(en_l)$ represents the software execution time.

The costs $c^a(v_j)$, $c^{th}(v_j)$, $c^{dm}(v_j)$, $c^{pm}(v_j)$ and $c^{ts}(v_j)$ for the instances v_j of entity en_l are equal to the costs of en_l. The interface costs for an edge $e = (v_1, v_2)$ are considered as follows:

1. $ci^a(e)$ defines the additional hardware area.
2. $ci^t(e)$ is the communication time for e.

A *design* represents the realization of a system S on a target architecture TA. The *design quality* can be expressed by the following *design metrics*.

1. $C^a(S)$ is the hardware area of S.
2. $C^{pm}(S)$ is the used software program memory of S.
3. $C^{dm}(S)$ is the used software data memory of S.
4. $C^t(S)$ is the total execution time of S.

The set of *design constraints, C,* consists of $MAX^a(S)$, $MAX^{pm}(S)$, $MAX^{dm}(S)$, and $MAX^t(S)$.

Definition 9.3

The **hardware–software partitioning problem** *is the problem of finding a mapping map:* $V \rightarrow TA$, *in such a way that all performance and resource constraints are fulfilled and the design costs are minimized.*

We need a few more notations to describe the IP model. In the following, we introduce these:

$J = \{1, \ldots, n_V\}$ represents the indices of $v_j \in V$.

$K = \{1, \ldots, n_P\}$ represents the indices of elements $ta_k \in TA$.

$L = \{1, \ldots, n_{EN}\}$ represents the indices of elements $en_l \in EN$.

$c_{l,k}^x$ be the cost metric $c^x(en_l)$ on target architecture component ta_k.

$c_{j,k}^x$ be the cost metric $c^x(v_j)$ on target architecture component ta_k.

C_k^x be the design metric $C^x(S)$ on ta_k of S.

MAX_k^x be the design constraint $MAX^x(S)$ on ta_k of S.

T_j^S be the execution starting time of node v_j.

T_j^D be the execution time of node v_j.

T_j^E be the execution ending time of node v_j.

The integer programming model requires the following 0/1-variables:

$$x_{j,0} = \begin{cases} 1 & : \quad v_j \text{ is executed unshared on } ta_0, \\ 0 & : \quad \text{otherwise.} \end{cases}$$

$$y_{j,k} = \begin{cases} 1 & : \quad v_j \text{ is executed shared on } ta_0, \\ 1 & : \quad v_j \text{ is executed on processor } ta_k(k \geq 1), \\ 0 & : \quad \text{otherwise.} \end{cases}$$

$$sh_{l,k} = \begin{cases} 1 & : \quad en_l \text{ is executed shared on } ta_0, \\ 1 & : \quad en_l \text{ is executed on processor } ta_k(k \geq 1), \\ 0 & : \quad \text{otherwise.} \end{cases}$$

$$i_{j_1,j_2} = \begin{cases} 1 & : \quad \text{an interface needed between } v_{j_1} \text{ and } v_{j_2}, \\ 0 & : \quad \text{otherwise.} \end{cases}$$

$$b_{j_1,j_2,k} = \begin{cases} 1 & : \quad v_{j_1} \text{ and } v_{j_2} \text{ are executed on different components,} \\ 1 & : \quad v_{j_1} \text{ ends before } v_{j_2} \text{ starts on } ta_k, \\ 0 & : \quad \text{otherwise.} \end{cases}$$

The set of constraints to be satisfied are as follows:

1. **General constraints:** Each node v_j is executed exactly on one target architecture component ta_k.

$$\forall j \in J : x_{j,0} + \sum_{k \in K} y_{j,k} = 1$$

2. **Resource constraints:** This is about the resources used by the system. On the software side, it is the total memory space required, while on the hardware side, it is the hardware

area. The values for used data memory C_k^{dm} and program memory C_k^{pm} on each processor ta_k should not exceed a given maximum. On the other hand, the used hardware area C_0^a is the sum of hardware area of unshared instances, shared entities and the total interface area CI_0^a. Again, C_0^a should not exceed a given maximum. That is,

$$\forall k \in K - \{0\} : C_k^{dm} = \sum_{l \in L} sh_{l,k} \times c_{l,k}^{dm} \leq MAX_k^{dm}$$

$$\forall k \in K - \{0\} : C_k^{pm} = \sum_{l \in L} sh_{l,k} \times c_{l,k}^{pm} \leq MAX_k^{pm}$$

$$C_0^a = \sum_{j \in J} x_{j,0} \times c_{j,0}^a + \sum_{l \in L} sh_{l,0} \times c_{l,0}^a + CI_0^a \leq MAX_0^a$$

3. **Timing constraints:** To determine the starting and ending times of each node, scheduling has to be performed. The execution time T_j^D of v_j is either the hardware or the software execution time. The ending time T_j^E is the sum of the starting time T_j^S and the execution time T_j^D. The starting time T_j^S of nodes have to be in their ASAP/ALAP range which can be considered for all edges $e = (v_{j_1}, v_{j_2})$ including interface communication time T_{j_1,j_2}^I. The system execution time C^t is the maximum of all ending times and thus must not violate the constraint. That is,

$$\forall j \in J : T_j^D = x_{j,0} \times c_{j,0}^{th} + y_{j,0} \times c_{j,0}^{th} + \sum_{k \in K - \{0\}} y_{j,k} \times c_{j,k}^{ts}$$

$$\forall j \in J : T_j^E = T_j^S + T_j^D$$

$$\forall j \in J : \quad ASAP(v_j) \leq T_j^S \leq ALAP(v_j)$$

$$\forall e = (v_{j_1}, v_{j_2}) \in E : T_{j_2}^S \geq T_{j_1}^E + T_{j_1,j_2}^I$$

$$\forall j \in J : T_j^E \leq C^t \leq MAX^t$$

An interface has to be realized for an edge $e = (v_{j_1}, v_{j_2})$, if v_{j_1} and v_{j_2} are realized in different target architecture components. Next, using the interface 0/1 variables i_{j_1,j_2}, the interface costs, like interface execution time T_{j_1,j_2}^I, interface hardware area A_{j_1,j_2}^I and the area of all interfaces CI_0^a can be calculated as follows. The costs are to be minimized.

$$\forall e = (v_{j_1}, v_{j_2}) \in E :$$

$$i_{j_1,j_2} \geq x_{j_1,0} + y_{j_1,0} + \sum_{k \in K - \{0\}} y_{j_2,k} - 1$$

$$i_{j_1,j_2} \geq x_{j_2,0} + y_{j_2,0} + \sum_{k \in K - \{0\}} y_{j_1,k} - 1$$

$$i_{j_1,j_2} \geq \sum_{k_1 \in K - \{0\}} \sum_{k_2 \in K - \{0\}, k_2 \neq k_1} y_{j_1,k_1} + y_{j_2,k_2} - 1$$

$$T_{j_1,j_2}^I = i_{j_1,j_2} \times ci_{j_1,j_2}^t$$

$$A_{j_1,j_2}^I = i_{j_1,j_2} \times ci_{j_1,j_2}^a$$

$$CI_0^a = \sum_{e = (v_{j_1}, v_{j_2}) \in E} A_{j_1,j_2}^I$$

Next, we consider the sharing of nodes on hardware and software. An entity en_l is shared on hardware ta_0 if at least two nodes v_{j_1} and v_{j_2} which are instances of en_l executed and shared on ta_0. An entity en_l is shared on processor ta_k if at least one instance of entity en_l is executed on ta_k.

$$\forall l \in L : \forall j_1, j_2 \in J : I(v_{j_1}) = I(v_{j_2}) = en_l \quad : \quad sh_{l,0} \geq y_{j_1,0} + y_{j_2,0} - 1$$

$$\forall k \in K - \{0\} : \forall l \in L : \forall j \in J : I(v_j) = en_l \quad : \quad sh_{l,k} \geq y_{j,k}$$

To consider the scheduling of nodes, it may be noted that two nodes v_{j_1} and v_{j_2} can be executed in parallel have to be sequentialized if,

- v_{j_1} and v_{j_2} are executed on the same processor, or
- v_{j_1} and v_{j_2} are shared on hardware.

To sequentialize two nodes v_{j_1} and v_{j_2} on a target architecture component ta_k, the binary decision variables $b_{j_1,j_2,k}$ and $b_{j_2,j_1,k}$ can be utilized as follows. The variables must have different values.

$$
\begin{aligned}
b_{j_1,j_2,k} + y_{j_1,k} &\geq 1 \\
b_{j_1,j_2,k} + y_{j_2,k} &\geq 1 \\
b_{j_1,j_2,k} + b_{j_2,j_1,k} &\geq 1 \\
b_{j_1,j_2,k} + b_{j_2,j_1,k} + y_{j_1,k} + y_{j_2,k} &\leq 3 \\
\forall k \in K : T_{j_1}^S &\geq T_{j_2}^E - BIGNUM \times b_{j_1,j_2,k} \\
T_{j_2}^S &\geq T_{j_1}^E - BIGNUM \times b_{j_2,j_1,k}
\end{aligned}
$$

where, $BIGNUM$ is a very big positive number.

Once these equations have been formulated, any standard IP-solver can be used to solve the particular instance. However, the basic difficulty of the approach is the exponentially increasing runtime of IP-solvers with increasing number of variables and constraints. Hence, the method can be used only for small task graphs.

9.2 Extended Kernighan-Lin Heuristic

Kernighan-Lin heuristic is a fast and efficient method to bi-partition a graph/hypergraph. It generates two partitions of equal size (same number of nodes). It is an iterative improvement-based algorithm, in the sense that it improves upon an initial partitioning of nodes. Primarily used in the domain of circuit partitioning to generate a module clustering such that intermodule wire length (and thus, delay) reduces, this has been applied in many other domains as well. The major advantage of the scheme is its extremely low runtime (of the order of few seconds for moderately sized graph), it consistently yields excellent results. Many modifications have also been suggested to produce imbalanced partitions (partitions having unequal number of nodes) and some efficient data structures to make the strategy more attractive.

The concept of bipartitioning can be naturally extended to the problem of hardware-software partitioning. The system specification can be considered as a task graph. Individual nodes are the tasks involved in the process. A node may either be implemented in hardware or in software. When implemented in hardware, a node has associated features like computation time, area, power, etc. On the other hand, for a software implementation also the features are computation time, memory required, power, etc. Any system specification can be considered to be consisting of a set of tasks, each forming a node. The nodes may have directed edges between them, identifying the tasks calling each other. For example, Fig. 9.1 shows a set of five interacting tasks $n1$, $n2$, $n3$, $n4$, $n5$. Each node (task) has an *internal computation time* (ict) needed to perform it in hardware or software. An edge, for example, $e1$ represents that the task $n1$ calls task $n2$ to get some operation done. Each edge has got with it attributes identifying the number of times the destination task is called by the source, and the amount of data transfer involved in each such call. Now, if the first task is implemented in software and the second one in hardware, or vice versa, the attributes associated with edges identify the amount of communication time needed in the implementation. The final implementation architecture is shown in Fig. 9.2.

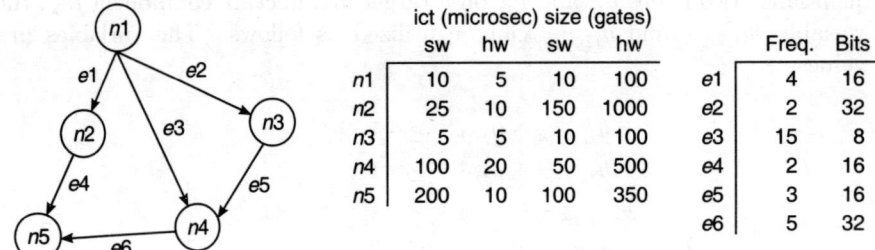

Fig. 9.1 Example task graph with associated attributes.

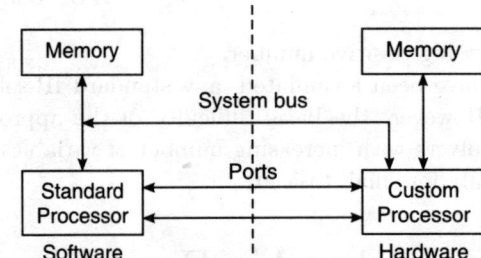

Fig. 9.2 An implementation architecture.

9.2.1 The Kernighan-Lin Heuristic

Given an initial bipartitioning of graph nodes, the Kernighan-Lin algorithm tries to improve it in an iterative fashion. The bipartitioning represents a cut of graph nodes. The cost of the cut is defined to be the sum of edge-costs of all the edges crossing the cut. The algorithm attempts to evolve better bipartitioning reducing the cut size.

At each iteration of the algorithm, first, all nodes are declared as *unlocked*. It then measures the costs of resulting cuts through all possible swaps of unlocked nodes in opposite parts. The swap resulting in the greatest cost decrease or least cost increase is taken and the pair is declared to be locked. The process continues till there are unlocked pairs. When all nodes are locked, the algorithm chooses the lowest cost partition seen, thus completing one iteration. If an iteration does not reduce the cost at all, the algorithm terminates. Otherwise, the process is repeated after unlocking all nodes.

Algorithm KL-heuristic
Input: A set of nodes $N = \{n_1, n_2, \ldots n_m\}$ partitioned into non-empty equal sized sets p_1
    and p_2, such that $p_1 \cup p_2 = N$, $p_1 \cap p_2 = \phi$.
Output: A refined partition with reduced cut size.
      do
            current-partition = best-partition = (p_1, p_2)
            Unlock all nodes
            while *unlocked-nodes-exist (current-partition)* do
                swap = Select-next-move (current-partition)
                current-partition = Move-and-lock-nodes (current-partition, swap)
                best-partition = Get-better-partition (best-partition, current-partition)
            end while
            if not *(Cost-fct(best-partition) < Cost-fct((p_1, p_2)))* then
                return (p_1, p_2) // Terminate, no improvement
            else // Do another iteration
                (p_1, p_2) = *best-partition*
                Unlock all nodes
            end if
      end do
procedure *Select-next-move (P)*
      for each unlocked $(n_i \in p_1, n_j \in p_2)$ do
            Append *(costlog, CostFct(Swap(P, n_i, n_j))*
      end for
      return $(n_i, n_j$ swap in costlog with lowest cost)

Figure 9.3 shows a graph with 8 nodes. All edge weights are assumed to be unity, for the sake of simplicity. We assume an initial bipartition of the graph with *partition-1* = {1, 2, 3, 4} and

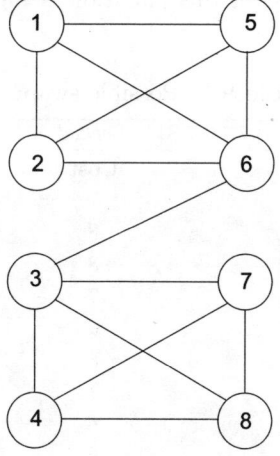

Fig. 9.3 An example graph for Kernighan-Lin partitioning.

partition-2 = {5, 6, 7, 8}. Thus, the cost of the partition (measured in terms of number of edges crossing the cut) is 9. In the first iteration of the *while* loop, we try out all possible swappings between the nodes of the partitions. The function *Select-next-move* generates the *costlog* as shown in Table 9.1.

Table 9.1 Possible swappings

$P = \{\{1,2,3,4\},\{5,6,7,8\}\}$, *Locked* = $\{\}$, *Cost* = 9	
Swap	Cost
1,5	9
1,6	8
1,7	7
1,8	7
2,5	9
2,6	8
2,7	7
2,8	7
3,5	**6**
3,6	7
3,7	8
3,8	8
4,5	7
4,6	**6**
4,7	9
4,8	9

Since swap (3,5) results in a minimum cost, the nodes are interchanged and are locked. The partition {1,2,4,5}, {3,6,7,8} is remembered as the *best-partition*. In the next iteration of the *while* loop, the swaps noted in Table 9.2 are computed. The nodes (4,6) are exchanged between the partitions and are locked resulting in current partitions {1,2,5,6} and {3,4,7,8}. The partition is remembered as *best-partition*. The *costlog* for the next iteration of the *while* loop is shown in Table 9.3. Nodes (1,7) are exchanged and locked, giving current partition {2,5,6,7}, {1,3,4,8}. However, cost of this partition is higher than the *best-partition*, hence *best-partition* is not affected.

Table 9.2 Possible swappings

$P = \{\{1,2,4,5\},\{3,6,7,8\}\}$, *Locked* = $\{3,5\}$, *Cost* = 6	
Swap	Cost
1,6	7
1,7	8
1,8	8
2,6	7
2,7	8
2,8	8
4,6	**1**
4,7	6
4,8	6

Table 9.3 Possible swappings

$P = \{\{1,2,5,6\},\{3,4,7,8\}\}$, $Locked = \{3,4,5,6\}$, $Cost = 1$	
Swap	Cost
1,7	7
1,8	7
2,7	7
2,8	7

In the next iteration of the *while* loop, we have only two unlocked nodes 2 and 8. Exchanging them results in the partition $\{1,2,3,4\}$, $\{5,6,7,8\}$ with cost 9. This does not affect the *best-partition*. All nodes are now locked. So, we come out of the *while* loop. The next *if* condition check fails as cost of *best-partition* $\{\{1,2,5,6\},\{3,4,7,8\}$ is 1, which is lesser than the given partition $\{p_1 = \{1,2,3,4\}, p_2 = \{5,6,7,8\}\}$. As a result, (p_1, p_2) becomes equal to the *best-partition*, all nodes are unlocked and the next iteration of the *do* loop starts. In this iteration also similar types of *costlog*s are generated, however *best-partition* never gets updated. The *while* loop finishes when all nodes get locked. This time, the *if* condition check is successful. The algorithm terminates with the partition $\{\{1,2,5,6\}, \{3,4,7,8\}\}$ having cost of 1.

9.2.2   Extending KL-Heuristic for Hardware–Software Partitioning

To apply KL heuristic to hardware–software partitioning, the following points are to be noted:

1. The objective of partitioning a task graph into hardware and software modules is to achieve better performance. From this angle, the cut-size metric is conveniently replaced by execution time metric. Execution time of node n, represented as $n.et$ (execution time) is given by,

$$n.et = n.ict + n.ct$$

where $n.ict$ is the internal computation time of node n. It may be noted that it can represent either the hardware time or the software time based upon the mapping of the node n to hardware or software, respectively. $n.ct$ encompasses the computation times of the tasks called by task n. Thus,

$$n.ct = \sum_{e_k \in n.outedges} e_k.freq \times (e_k.tt + (e_k.dst).et)$$

$e_k.tt$ represents the transfer time for edge e_k. An expression for $e_k.tt$ can be written as,

$$e_k.tt = \lceil busdelay \times (e_k.bits/buswidth) \rceil$$

For example, for the task graph shown in Fig. 9.1, we have the following execution times for individual nodes:

$$
\begin{aligned}
n_5.et &= n_5.ict \\
n_4.et &= n_4.ict + e_6.freq \times (e_6.tt + n_5.et) \\
&= n_4.ict + e_6.freq \times (e_6.tt + n_5.ict)
\end{aligned}
$$

$$
\begin{aligned}
n_3.et &= n_3.ict + e_5.freq \times (e_5.tt + n_4.et) \\
&= n_3.ict + e_5.freq \times (e_5.tt + n_4.ict + e_6.freq \times (e_6.tt + n_5.ict)) \\
n_2.et &= n_2.ict + e_4.freq \times (e_4.tt + n_5.et) \\
&= n_2.ict + e_4.freq \times (e_4.tt + n_5.ict) \\
n_1.et &= n_1.ict + e_1.freq \times (e_1.tt + n_2.et) + e_2.freq \times (e_2.tt + n_3.et) \\
&\quad + e_3.freq \times (e_3.tt + n_4.et) \\
&= n_1.ict + e_1.freq \times (e_1.tt + n_2.ict + e_4.freq \times (e_4.tt + n_5.ict)) \\
&\quad + e_2.freq \times (e_2.tt + n_3.ict + e_5.freq \times (e_5.tt + n_4.ict \\
&\quad + e_6.freq \times (e_6.tt + n_5.ict))) \\
&\quad + e_3.freq \times (e_3.tt + n_4.ict + e_6.freq \times (e_6.tt + n_5.ict))
\end{aligned}
$$

In this example, since n_1 is the root node of the task graph, minimizing $n_1.et$ will reduce the overall time required. It may be noted that the model has some inaccuracies since some computation and communication can occur in parallel. Also, the bus loading factor has been ignored which can affect the transfer time.

2. Redefine a move as a single object move. Basic Kernighan-Lin algorithm assumes a balanced partition. Each move is a swap of a pair or nodes belonging to two partitions. Thus, the partition sizes are always balanced. In HW-SW partitioning, although we have two parts, software and hardware, the units of sizes for the two parts are different. For software modules, it is the memory required to store the instructions, whereas, for the hardware part, it may be the number of gates needed. Thus, having a balanced partition is not really meaningful here. Thus, a move is defined to be a single object moving from hardware partition to software, or vice versa.

3. Use better data structure for faster move selection. In the Kernighan-Lin heuristic, to select the move resulting in maximum gain, we need to try out all possible moves and then select from them. This involves good amount of computation. However, not all quantities in the equations corresponding to execution times of nodes change when any node changes its partition—only a few quantities can be recomputed to get the new execution times. For example, if node n_1 changes its partition, the quantities affected are $n_1.ict$, $e_1.tt$, $e_2.tt$, and $e_3.tt$. If these quantities change by amounts $Dn_1.ict$, $De_1.tt$, $De_2.tt$ and $De_3.tt$, then the change in n_1's execution time, $Dn_1.et$ is given by,

$$
Dn_1.et = Dn_1.ict + De_1.tt \times e_1.freq + De_2.tt \times e_2.freq + De_3.tt \times e_3.freq
$$

Similar change equations can be derived for other node movements as follows:

For movement of node n_2,

$$
Dn_1.et = e_1.freq \times (De_1.tt + Dn_2.ict + e_4.freq \times De_4.tt)
$$

For movement of node n_3,

$$
Dn_1.et = e_2.freq \times (De_2.tt + Dn_3.ict + e_5.freq \times De_5.tt)
$$

For movement of node n_4,

$$
Dn_1.et = e_2.freq \times e_5.freq \times (De_5.tt + Dn_4.ict + e_6.freq \times De_6.tt)
$$
$$
+ e_3.freq \times (De_3.tt + Dn_4.ict + e_6.freq \times De_6.tt)
$$

For movement of node n_5,

$$Dn_1.et = e_1.freq \times e_4.freq \times (De_4.tt + Dn_5.ict)$$
$$+ e_2.freq \times e_5.freq \times e_6.freq \times (De_6.tt + Dn_5.ict)$$
$$+ e_3.freq \times e_6.freq \times (De_6.tt + Dn_5.ict)$$

Let *Max-cost* be the maximum cost (computation time) for the whole system resulting from the worst case implementation of the system. If we maintain an array, *Change-list* with index running from $-Max$-$cost$ to $+Max$-$cost$, then each of these quantities $Dn_1.et$ for the movement of n_1, n_2, n_3, n_4 and n_5 corresponds to an index in this array. The index closest to $-Max$-$cost$ is the best movement as this represents the maximum gain.

If all the tasks are initially assumed to be in software, n_1's execution time,

$$n_1.et = 10 + 4 * (25 + 2 * 200) + 2 * (5 + 3 * (100 + 5 * 200)) + 15 * (100 + 5 * 200)$$
$$= 24820 \ \mu s$$

If we assume that the bus is an 8-bit one and each transfer over the bus takes 2 μs, then the transfer times (in μs) for the edges are as given below. It may be noted that these transfer times come into existence only when the source and destination tasks of the edges lie on different partitions, that is, one in hardware and the other in software. For hardware–hardware and software–software communications, the transfer times are negligible and are assumed to be zero.

$$e_1.tt = 2 * 16/8 = 4$$
$$e_2.tt = 2 * 32/8 = 8$$
$$e_3.tt = 2 * 8/8 = 2$$
$$e_4.tt = 2 * 16/8 = 4$$
$$e_5.tt = 2 * 16/8 = 4$$
$$e_6.tt = 2 * 32/8 = 8$$

Now, if we compute $Dn_1.et$ values for movement of different nodes we get the quantities noted in Table 9.4. Thus, the array *Change-list* will be as shown in Fig. 9.4. Thus, node n_5 moves to hardware. $n_1.et$ becomes 4222. Since nodes n_2 and n_4 call node n_5, the corresponding change equations need to be recomputed. Change equations for n_1 and n_3 are unaffected and the node n_5 is locked. Thus, the new $Dn_1.et$ quantities are as shown in Table 9.5. As a consequence, node n_4 moves to hardware. $n_1.et$ becomes 1756.

Table 9.4 $Dn_1.et$ for node movements

Moved node	$Dn_1.et$
n_1	57
n_2	−12
n_3	40
n_4	−456
n_5	−20598

Change equations for nodes n_1 and n_3 are to be recomputed as shown in Table 9.6. Thus, n_2 moves to hardware. $n_1.et$ becomes 1680. Table 9.7 shows the new situation. Thus, n_1

moves to hardware, $n_1.et = 1645$. Now, if n_3 moves to hardware, $Dn_1.et = -40$. Thus, the move is accepted and $n_1.et$ becomes 1605, which is same if all nodes are mapped to hardware.

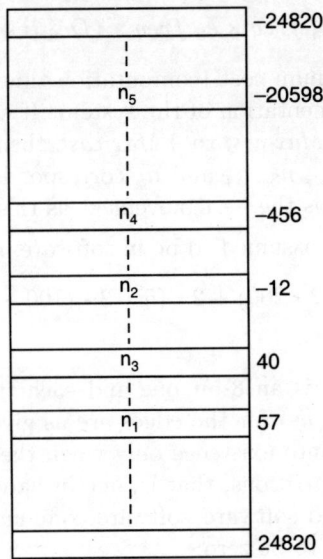

Fig. 9.4 Change-list for the movement of nodes.

Table 9.5 $Dn_1.et$ for node movements

Moved node	$Dn_1.et$	Comment
n_1	57	Unchanged
n_2	−76	$De_4.tt = -4$, as n_2 moves to hardware, this communication cost reduces
n_3	40	Unchanged
n_4	−2466	$De_6.tt = -8$

Table 9.6 $Dn_1.et$ for node movements

Moved node	$Dn_1.et$
n_1	−3
n_2	−76
n_3	−8

Table 9.7 $Dn_1.et$ for node movements

Moved node	$Dn_1.et$
n_1	−35
n_3	−8

This finishes one iteration of the heuristic. All nodes will be unlocked now and the next iteration can start to determine a better solution. It may be noted that some other cost metric, like area alone, or a combined area and execution time may yield different types of results.

9.3 Partitioning Using Genetic Algorithm

Genetic Algorithm (GA) is a stochastic search algorithm based on operations of natural genetics. Here, a fixed-sized population of chromosomes evolves over a number of generations following the principle of natural selection. Each chromosome identifies a potential solution. A chromosome has got an associated fitness measure. Using the operators similar to crossover and mutation in nature, the population evolves through generations.

In the hardware–software partitioning problem, a chromosome is a bit-string. If there are n nodes in the task graph to be partitioned, a chromosome is also taken to be n-bit long, with the ith bit identifying the partition for the ith node. It can be conveniently assumed to be '0' if the node is put into the hardware partition and '1' for software (or vice versa). A fixed number of such chromosomes can be created randomly to represent the initial population.

As the software and hardware cost metrics (such as area, time delay, power) of all nodes are known, for a chromosome, we can evaluate its fitness by considering the overall delay, hardware area, power etc.

To evolve a new generation, the top small percentage of chromosomes are directly copied to the next generation. Rest of the population is created by two operators—crossover and mutation. The crossover operator selects two parent chromosomes to participate in the operation. Their parts are exchanged to create new offspring. The crossover may be single-point, multi-point, etc. The mutation operator may be implemented by selecting a parent chromosome and randomly complementing some of its bits (i.e., changing partitions). The mutation rate can be controlled to control the rate of convergency to local or global minima.

The main drawback of such genetic approach is the slow rate of convergence. It often requires the GA to evolve a large number of generations to converge to a solution. The termination criteria is often set to be "no improvement in last few generations" or a maximum number of generations for which the GA has run. The best solution at the end is taken to be the solution of the partitioning problem. To accelerate the rate of convergence, the mutation rate can be increased, however, it mostly converges to local best solutions, rather than finding the global best.

9.4 Partitioning Using Particle Swarm Optimization (PSO)

Particle Swarm Optimization (PSO) is a population based stochastic technique developed by Eberhart and Kennedy in 1995. The PSO algorithm is inspired by social behaviour of bird flocking or fish schooling. In PSO, the potential solutions, called *particles*, fly through the problem space by following the current optimum particles. PSO has been successfully applied in many areas and has been found to outperform GA in cost function and execution time.

Consider the following scenario: A group of birds are randomly searching for food in an area. There is only one piece of food in the area being searched. None of the birds know exactly where the food is. During each iteration, they learn via their inter-communications, how far the food is. So the best strategy to find where the food is to follow the bird which is nearest to the food.

PSO is learnt from this bird-flocking scenario, and used to solve the optimization problems. Each single solution (particle) is perceived as a *bird* in the search space. Each particle has a fitness value which is evaluated by the fitness function (the cost function to be optimized), and has a velocity which directs its flight. The particles fly through the problem space by following the current optimum particles.

PSO is initialized with a group of random particles (solutions) and then searches for optima by updating through generations. During every generation (iteration), each particle is updated by following two *best* values. The first one is the position vector of the best solution (fitness) this particle has achieved so far. This fitness value is also stored along with the particle. This position is called *pbest*. Another best position that is tracked by the particle swarm optimizer is the global best position, obtained so far, by any particle in the population. This best position is the current global best and is called *gbest*.

After finding the two best values, the particle updates its velocity and position according to the following two equations.

$$v^i_{k+1} = wv^i_k + c_1 r_1 (pbest^i - x^i_k) + c_2 r_2 (gbest_k - x^i_k)$$
$$x^i_{k+1} = x^i_k + v^i_{k+1}$$

where v^i_{k+1} is the velocity of particle i at $(k+1)$th iteration, x^i_k is the current particle solution (position), r_1 and r_2 are random numbers between 0 and 1, c_1 is the self-confidence (cognitive) factor, c_2 is the swarm confidence (social) factor, w is the inertia factor.

The first term in the first equation represents the effect of inertia of the particle, the second term represents the particle memory influence and the third term represents the swarm influence. The velocities of the particles on each dimension may be clamped to a maximum velocity V_{max}, which is a parameter specified by the user. If the sum of accelerations causes the velocity on that dimension to exceed V_{max}, then it is limited to V_{max}. A comparison with GA shows that PSO is better than GA as it has lesser number of tuning parameters, and is faster than GA due to the linear complexity of its main loop.

Next, we will consider the PSO formulation of the hardware–software partitioning problem. Consider a design that consists of m nodes. A possible solution (particle) is a vector of m elements, where each element is associated to a given node. The element assumes a '0' value if it is implemented in hardware and a '1' value if it is implemented in software. There are n particles (solutions). The particles are initialized randomly. Each node of the design has got associated cost parameters—the hardware implementation cost (HW_{cost}), the software implementation cost (SW_{cost}), the power cost (POW_{cost}). The communication cost is assumed to be absorbed within the hardware and software implementation costs. The cost of a particle is then evaluated as,

$$\text{Cost} = HW_{cost} \, / \, allHW_{cost} + SW_{cost} \, / \, allSW_{cost} + POW_{cost} \, / \, allPOW_{cost}$$

where $allHW_{cost}$ is the maximum hardware cost when all the nodes are mapped to hardware and so on.

The velocity of each node is initialized in the range of -1 to 1, where a negative velocity indicates that the particle is moving towards 0 and a positive velocity indicates that the particle is moving towards 1. In each iteration of the algorithm, the equations noted earlier for velocity and position calculation are evaluated. If the particle goes out of the permissible region, it is clamped to the nearest limit. In particular, as per these equations, the particle node values can take any real value between 0 and 1. But as a binary problem, the node values must be rounded to 0 or 1. Therefore, the position has to rounded to the nearest integer (0 or 1). The algorithm terminates if the improvement in the global best solution, *gbest* remains less than a predefined value (ϵ) for a predefined number of iterations.

9.5   Extended Partitioning Problem

In the previous sections, we have seen a few strategies to perform hardware–software partitioning. However, an important aspect of system design is the availability of multiple design options for every task in a task-graph. For example, even if a node has been mapped to hardware, there may be several implementations possible for it. Same can be true for software modules as well. Thus, it is not sufficient just to determine whether a node is to be mapped to hardware or software—the appropriate implementation needs to be determined as well. Another important aspect of partitioning is scheduling, i.e., determining which node executes at which time. If the decision of scheduling is deferred till the finalization of the partitioning and the implementation of alternative selection, it leaves little room for the scheduling procedure to meet the timing constraints. Hence, scheduling should also be done simultaneously with mapping and implementation selection. Thus, the goal of partitioning is to determine three parameters for each task: mapping (hardware or software), implementation (type or realization), and schedule. However, since the overall problem is quite complex, we will look into it in two stages—*binary partitioning* and *extended partitioning*. The binary partitioning is to determine for each node, a hardware or a software mapping and the associated schedule. It does not consider the availability of multiple implementation alternatives within hardware or software. We will look into this problem first. The extended partitioning considers the selection of an appropriate implementation for each node, over and above the binary partitioning. However, before discussing either of the two problems, we consider a set of assumptions that will be utilized in the formulation.

1. The tasks form a task-graph. The task-graph is a directed acyclic graph (DAG) with a deadline D clock cycles for the completion of all tasks.

2. The target architecture consists of a single programmable processor (for software execution) and a custom datapath (the hardware component). The software size should not exceed AS (memory capacity) and the hardware size should not exceed AH (area, gate-count, etc.). The communication costs of hardware–software interface are represented by,

ah_{comm}:   hardware area required to communicate one sample of
data across the interface

as_{comm}:   software area required to communicate one sample of
data across the interface

t_{comm}:   number of cycles required to transfer the data

3. The area and time estimates corresponding to each of the hardware and software implementation of alternatives for each of the task nodes are known. For node i, CH_i and CS_i are the hardware and software implementation curves. A curve plots all possible design alternatives (implementation bins). Thus, $CH_i = \{(ah_i^j, th_i^j), j \in NH_i\}$, where ah_i^j and th_i^j represent the area and execution time when node i is implemented in hardware bin j, and NH_i is the set of all hardware bins. Similarly, $CS_i = \{(as_i^j, ts_i^j), j \in NS_i\}$, where,

as_i^j and ts_i^j represent the program size and execution time when node i is implemented in software bin j. Within a mapping (hardware or software), the fastest implementation bin is called L bin and the slowest implementation bin is called the H bin. For binary partitioning, of course, it is to be considered that for each node, there is exactly one software implementation bin and also a single hardware implementation bin.

4. Reuse between the nodes mapped to hardware is not considered.

9.5.1 Binary Partitioning

In this section, we will discuss the solution to the binary partitioning problem. The algorithm is named as, *Global Criticality/Local Phase* (*GCLP*) algorithm. The binary partitioning can be stated as follows:

Given a DAG, area and time estimates for software and hardware mappings of all nodes, and communication costs, subject to resource capacity constraints and a deadline D, determine for each node i, the hardware or software mapping (M_i) and the start time for the execution of the node (schedule) such that the total area occupied by the nodes mapped to hardware is at its minimum.

The *GCLP* algorithm is built around the principle of *list scheduling* used to schedule nodes in a DAG. Based upon some priority function, the nodes are serialized. One possible priority function may be their order of occurrences in the DAG. The list is traversed serially and for each node, a mapping that minimizes an objective function is selected. From the context of hardware–software partitioning, at each node, one of the following two objective functions can be used:

1. Minimize the finish time of the node.
2. Minimize the area (hardware or software) of the node.

However, none of these two objectives can target the solution of the binary partitioning problem as a whole, since it has to minimize area and meet timing constraints simultaneously for the overall DAG. The objective of reducing finish time of a node may lead to the satisfaction of deadline constraint, however, may result in a highly suboptimal result from the point of view of area minimization. On the other hand, if the node mapping is always towards the minimum area implementation, it may violate the timing constraint for the DAG. Thus, a fixed objective function is unable to solve the constrained optimization problem. Moreover, the list scheduling has a limitation in the sense that mapping based on serial traversal tends to be greedy, thus globally suboptimal.

The GCLP algorithm adaptively selects an appropriate *mapping objective* at every step to determine the mapping of a node and its schedule. The mapping objective procedure has been illustrated in Fig. 9.5. It is based on two measures:

1. *Global Criticality* (GC): This is a look-ahead measure that estimates the time criticality at each step of the algorithm. *GC* is compared with a threshold value. If *GC* is higher than the threshold (i.e., time is critical), the objective function that minimizes the finish time is selected. Otherwise, the objective function that minimizes area is selected.

2. *Local Phase* (LP): This is a classification of nodes based on their heterogeneity and other intrinsic properties. A node is classified to be in one of the three classes—*local phase 1 (extremity)*, *local phase 2 (repeller)*, or *local phase 3 (normal)*. A measure called *local*

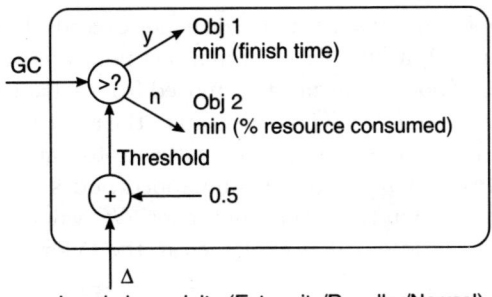

Fig. 9.5 Selection of mapping objective in GCLP.

phase delta quantifies the local mapping preferences of a node, and thus modifies the threshold value in GC comparison.

The GCLP algorithm proceeds as shown in **Algorithm GCLP** noted below. Let, N be the set of nodes in the DAG, N_M be the set of mapped nodes up to the current step and N_U be the set of unmapped nodes remaining. At each step of the algorithm, the global criticality value is estimated based on the currently mapped and unmapped nodes and the deadline requirements. A node is selected from the set of unmapped nodes whose predecessors have already been mapped. The node selection is based on an urgency criterion, a node on the critical path of the design is selected first. The local phase for the node is identified and the value of delta is computed. The GC and local phase delta values are now compared to identify the mapping objective for this step.

Algorithm overview-GCLP
Begin
    $N_U = N$, $N_M = \phi$
    while $N_U \neq \phi$ do
        Compute GC
        $i \leftarrow$ Select node among ready nodes (predecessors already mapped)
        Identify local phase and compute delta
        Determine objective function based on Fig. 9.5
        Select mapping for node i
        Find start time t_i for node i
        $N_M \leftarrow N_M \cup \{i\}$
        $N_U \leftarrow N_U - \{i\}$
        Update $T_{remaining}$
    end while
End.

Global Criticality Computation

As noted earlier, GC is a global look-ahead measure to identify the time criticality at each step of the algorithm. At any point of GCLP algorithm execution, a number of nodes are already mapped and others are yet to be mapped. Let, T_{rem} be the remaining time considering the deadline D and the finish times of mapped nodes. Let T^S be the finish time when all the remaining nodes are mapped to software. If T^S does not exceed D, the criticality value is less. However, if it is more than D, a number of unmapped nodes must be mapped in hardware. Let, $N_{S \to H}$ denote the set of nodes that must be moved from software to hardware to meet the deadline. The new finish time is T^H. GC at this step of the algorithm is defined as the fraction of unmapped nodes that have to be moved from software to hardware to meet feasibility. A high value of GC indicates that many as-yet unmapped nodes should go to hardware, thus time is a critical resource here. On the other hand, a low GC value means we can still continue with software-based implementation. It is explained in the following procedure.

Procedure Compute_GC
Begin
   Step 1: $N_{S \to H} \leftarrow \phi$
   Step 2: Select a node i from N_U using priority function Pf to be removed to hardware to meet deadline D.
   Step 3: $N_{S \to H} \leftarrow N_{S \to H} \cup \{i\}$
   Step 4: Compute T^H, finish time with nodes in $N_{S \to H}$ mapped to hardware.
   Step 5: If $T^H > D$ goto Step 2.
   Step 6: $GC = \frac{\sum_{i \in N_{S \to H}} size_i}{\sum_{i \in N_U} size_i}$, $0 \le GC \le 1$. The size of a node is taken to be the number of elementary operations (add, multiply, etc.) in the node.
End.

The priority function Pf used in Step 2 of the algorithm can be any of the following:

- Rank the nodes in decreasing order of software execution times.
- Use (ts_i/th_i) as the ranking function. That is, move nodes with high relative gain in time while moved from software to hardware.
- Rank nodes based on ah_i, increasing hardware area.

Experimentation shows the second priority function to work well for most of the cases.

Local Phase (LP) Computation

The global criticality (GC) is an average measure over all unmapped nodes identifying whether the mapping be guided towards reducing the finish time or the resource requirement. However, the next node selected for mapping is a task which may or may not have very good hardware and/or software implementation. Thus, we need to consider the property of this node locally and appropriate threshold value may be set for GC to be compared with. The nodes can be classified into one of the following three categories:

- Extremities—local phase 1 nodes
- Repellers—local phase 2 nodes
- Normal—local phase 3 nodes

Local Phase 1: Extremity Nodes
These are the nodes that are best suited either in hardware or in software. The bottleneck resource in hardware is the area, while the bottleneck resource for software is the execution time. Extremities are nodes that consume a disproportionately large amount of bottleneck resource. A *hardware extremity* node consumes a large area in hardware, but a relatively small

amount of time in software. A *software extremity* node is just the reverse, requiring a large amount of time in software, but relatively small amount of area in hardware. The disparity in the resources consumed by an extremity node i is quantified by the *extremity measure*, E_i. It is computed as follows:

Procedure Compute_Extremity_Measure

Input: $ts_i, ah_i, \forall i \in N, \alpha, \beta$ percentiles

Output: $E_i, \forall i \in N, -0.5 \leq E_i \leq 0.5$

Begin

Compute the histograms of all nodes with respect to their software execution times (ts_i) and hardware areas (ah_i).

Determine $ts(\alpha)$ and $ah(\beta)$ corresponding to α and β percentiles of ts and ah histograms, respectively.

For each node i

If $ts_i \geq ts(\alpha)$ and $ah_i < ah(\beta)$ then, classify i to be software extremity, $i \in EX_s$

If $ah_i \geq ah(\beta)$ and $ts_i < ts(\alpha)$ then, classify i to be hardware extremity, $i \in EX_h$

If $i \in EX_s$, set extremity value $x_i = \frac{ts_i/ts_{max}}{ah_i/ah_{max}}$ else $x_i = \frac{ah_i/ah_{max}}{ts_i/ts_{max}}$,

where $ts_{max} = max_i\{ts_i\}$ and $ah_{max} = max_i\{ah_i\}$

Order the nodes in EX_s and EX_h by x.

Compute the extremity measure of node i as

If $i \in EX_s$ then $EX_i = -0.5 \times \frac{x_i - xs_{min}}{xs_{max} - xs_{min}}, -0.5 \leq E_i \leq 0$

else if $i \in EX_h$ then $EX_i = 0.5 \times \frac{x_i - xh_{min}}{xh_{max} - xh_{min}}, 0 \leq E_i \leq 0.5$

End.

First a distribution of the nodes with respect to their software execution time ts_i and hardware areas ah_i is computed. Parameters α and β represent the percentile cut-offs of these distributions. Typical values of these are in the range of 0.5 to 0.75. A node is classified as software extremity node if its execution time is beyond $ts(\alpha)$ and area is lesser than $ah(\beta)$. Similarly, a node is classified as hardware extremity if it is beyond the β percentile in area and below α percentile in software time. The extremity measure E_i is used to modify the default threshold in the direction of the preferred mapping. The new threshold is $0.5 + E_i$. The GC value is compared with this threshold to take the decision. For software extremities, $-0.5 \leq E_i \leq 0$, thus $0 \leq threshold \leq 0.5$, and for hardware extremities, $0 \leq E_i \leq 0.5$, thus, $0.5 \leq threshold \leq 1$.

Local Phase 2: Repeller Nodes

Nodes often contain intrinsic properties that reflect the inherent suitability of the node to hardware or software mapping. For example, bit operations are handled better in hardware than in software, whereas, memory operations are handled better in software. Thus, a node with many bit manipulations, relative to other nodes, is a software repeller. Thus, bit manipulation operation is a software repeller property. A repeller property is quantified by a *repeller value*. The combined effect of all repeller properties in a node is the *repeller measure* of the node. Given two nodes N_1 and N_2 with similar software characteristics, if N_1 has a higher software repeller measure than N_2, and given the choice of mapping one of them to hardware, N_1 is preferred.

Repeller Measure   The repeller measure R_i is a convex combination of all the repeller property values of a node. It is used to modify the threshold against which GC is compared. Let RH be the set of hardware repeller properties and RS be the set of software repeller properties. Let $P = RH \cup RS$ be the complete set of repeller properties.

> ### Procedure Compute_Repeller_Measure
> Input: $v_{i,p}$ = value of repeller property p for node i
> Output: Repeller measure R_i
> Begin
>
>>   For each property p compute:
>>> $\sigma^2(v_{i,p})$ = variance of $v_{i,p}$ over all i
>>> $min(v_{i,p})$ minimum of $v_{i,p}$ over all i
>>> $max(v_{i,p})$ maximum of $v_{i,p}$ over all i
>>>   Let $RX = RH$ if $p \in RH$ and $RX = RS$ if $i \in RS$.
>>>   Compute $a_p = \dfrac{\sigma^2(v_{i,p})}{\sum_{p \in RX} \sigma^2(v_{i,p})}$ = weight of repeller property p, $\sum_{p \in RX} a_p = 1$.
>>   Compute the normalized property value $nv_{i,p}$ for each property p of node i
>>> $nv_{i,p} = \dfrac{v_{i,p} - min(v_{i,p})}{max(v_{i,p}) - min(v_{i,p})}$, $0 \le nv_{i,p} \le 1$
>>   Compute the repeller measure R_i for node i as,
>>> $R_i = \frac{1}{2}(\sum_{p \in RH} a_p.nv_{i,p} - \sum_{p \in RS} a_p.nv_{i,p})$, $-0.5 \le R_i \le 0.5$.
> End.

The value of $v_{i,p}$ of each property p is identified by analysing the node. For example, for the bit-level instruction mix property, the number of bit-level operations in a node may be divided by the total number of operations for that node. The repeller measure R_i is used to modify the threshold so that the new threshold is $0.5 + R_i$. For software repellers, $-0.5 \le R_i \le 0$, so that $0 \le threshold \le 0.5$. For hardware repellers, $0 \le R_i \le 0.5$, so that $0.5 \le threshold \le 1$.

Local Phase 3: Normal Nodes
A node that is neither an extremity nor a repeller is a local phase 3 node or a normal node. The threshold value is set to be 0.5 in this case. The mapping of the node is governed by GC alone.

GCLP Algorithm

The following is the complete GCLP algorithm:

> ### Algorithm GCLP
> Input: ah_i, as_i, th_i, ts_i, E_i (extremity measure), R_i (repeller measure) $\forall i \in N$
>>   Communication costs: ah_{comm}, as_{comm}, and t_{comm}
>>   Constraints: AH, AS, D
> Output: Mapping M_i ($M_i \in$ {hardware, software}), start time t_i, $\forall i \in N$.

Begin

$N_U = \{unmapped\ nodes\} = N$, $N_M = \{mapped\ nodes\} = \phi$.

Compute GC.

Compute the effective execution time $t_{exec}(i)$ for each node i as follows

If $i \in N_U$, $t_{exec}(i) = GC \cdot th_i + (1 - GC) \cdot ts_i$

else $t_{exec}(i) = th_i$, if $M_i = $ hardware

$t_{exec}(i) = ts_i$, if $M_i = $ software

Compute the longest path $longestPath(i)$, $\forall i \in N_R$, the set of ready nodes, using $t_{exec}(i)$.

Select node i, $i \in N_R$, for mapping: $\max(longestPath(i))$

Determine mapping M_i for i as follows:

if $(E_i \neq 0)$ $\Delta = \gamma \cdot E_i$ (local phase 1)

where γ is extremity measure weight, $0 \leq \gamma \leq 1$

else if $(R_i \neq 0)$ $\Delta = v \cdot R_i$ (local phase 2)

where v is the repeller measure weight, $0 \leq v \leq 1$

else $\Delta = 0$ (local phase 3)

$Threshold = 0.5 + \Delta$, $0 \leq Threshold \leq 1$

If $GC \geq Threshold$ then $m : minimize(Obj1)$

else $m : minimize(Obj2)$

$M_i = m$; Set(t_i); $N_U = N_U - \{i\}$; $N_M = N_M \cup \{i\}$

Update $T_{remaining}$, $AH_{remaining}$, $AS_{remaining}$

The objective functions $Obj1$ and $Obj2$ are as follows:

Obj1: It selects a mapping that minimizes the finish time of a node. A node can begin execution only after all its predecessors have finished execution and the data has been transferred to it from the predecessors. For a node mapped onto software, execution cannot start until the last node mapped on software has finished. Thus, if $t_{fin}(i, m)$, where $m \in \{software, hardware\}$, be the finish time of node i with mapping m, it is given by the following expression:

$$t_{fin}(i, m) = max(max_{P(i)}(t_{fin}(p) + t_c(p, i)), tf_{last}(m)) + t(i, m)$$

where, $P(i)$ = set of predecessors of node i, $p \in P(i)$

$t_{fin}(p)$ = finish time of predecessor p

$t_c(p, i)$ = communication time between predecessor p and node i

$tf_{last}(m)$= finish time of last node assigned to mapping m

= 0 if m is hardware

$t(i, m)$ = execution time of node i on mapping m

Obj2: It uses a percentage resource consumption measure. It is the fraction of the resource area of a node (nodal + communication area) to the total resource area. It is defined as follows. It favours software allocation as the algorithm proceeds.

$$\frac{(as_i + as_{comm}^{tot})}{AS_{remaining}} \text{ if } m = software$$

$$\frac{(ah_i + ah_{comm}^{tot})}{AH_{remaining}} \text{ if } m = hardware$$

9.5.2   Extended Partitioning

The GCLP algorithm presented above solves the binary partitioning problem, in which, for a task node, there exists single software and hardware implementations. The extended partitioning problem couples with it the problem of selecting the implementation bin. A module may have different hardware realizations, called implementation bins. If L be the fastest implementation bin and H be the slowest one, as we traverse through the implementation bins from L towards H, the hardware area required decreases, while the delay increases. From the viewpoint of minimizing the hardware area, we may like to put all nodes in H-bin, however, that may be infeasible from the point of view of speed of the resulting system.

The basic algorithm proceeds as shown in Fig. 9.6. Initially, all nodes are marked as *free*. Assuming median area and time values for hardware and software implementations, the GCLP algorithm is applied to get mapping and schedule of all free nodes. A particular free node is then selected. The node becomes a *tagged node*. Assuming its mapping to be as given by the GCLP algorithm, an appropriate implementation bin is selected for the tagged node. The node thus becomes a *fixed node*. GCLP algorithm is then applied on the remaining nodes and the process is repeated until all nodes become *fixed*.

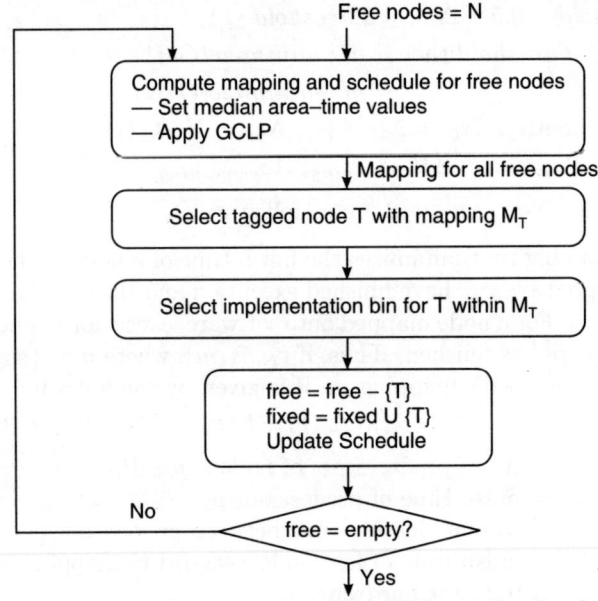

Fig. 9.6 Extended partitioning overview.

Implementation-bin Selection

Next, we present the procedure to select an implementation bin for the tagged node. We do it assuming that the node has been tagged to hardware, however, similar analysis can be made for nodes tagged to software as well. Let $free^h$ be the set of nodes marked to be implemented in hardware by the GCLP algorithm. All such nodes are assumed to be in their H (least area) bins initially. To select the implementation bin for the tagged node, the *Bin Fraction Curve* for the node is computed. The *bin fraction* BF_T^j computes for the tagged node T and for each of the implementation bin j, the fraction of $free^h$ nodes that need to be moved from H bins to the L bins, in order to meet the timing constraints. A high value of BF_T^j indicates that if the node T is implemented in bin j, a large number of nodes need to move to L bins, leading to significant increase in hardware area. The *bin fraction curve* (BFC_T) is the collection of all bin fraction values for the tagged node T. *Bin sensitivity* is the gradient of the bin fraction curve. If the maximum slope of the BFC lies between the bins $k-1$ and k, then, moving the tagged node from bin $k-1$ to k shifts the largest fraction of $free^h$ nodes to their L bins. That is, selecting bin $k-1$ instead of k results in largest reduction in hardware area for the $free^h$ nodes. Hence, $k-1$ bin is selected for implementation of the node.

An example of bin fraction curve and bin sensitivity has been shown in Fig. 9.7. Here, for the minimum area implementation of tagged node T, the bin selected should be H_T. However, it necessitates a large number of $free^h$ nodes to go to their L bins (as the BF_H^T value is quite high), increasing the overall hardware area. As we go down the BFC curve, the bin sensitivity is maximum at bin B_T^*. So, this bin is selected. If there are more than one BS_{max} values possible, as shown in Fig. 9.8, a *weighted bin sensitivity* is calculated. It is equal to the bin sensitivity multiplied by ah_T^H/ah_T^j. The selected bin is that one with maximum weighted bin sensitivity.

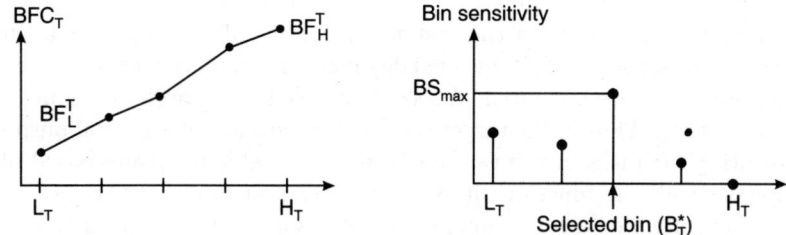

Fig. 9.7 An example bin fraction curve and bin sensitivity.

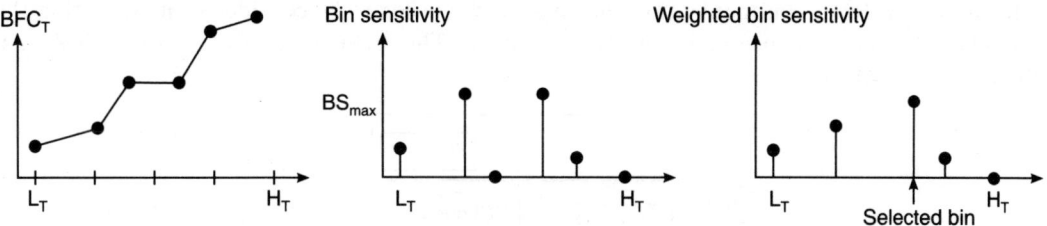

Fig. 9.8 Weighted bin sensitivity.

The overall extended partitioning algorithm is given next.

Algorithm Extended_Partitioning

Input: $\forall i \in N : CH_i, CS_i, E_i, R_i$

Interface communication costs: $ah_{comm}, as_{comm}, t_{comm}$

Constraints: AH, AS, and D

Output: $\forall i \in N$: mapping M_i (software or hardware), implementation bin B_i^*, start time t_i

Begin

$N_{fixed} = \phi$, $N_{free} = \{N\}$

Compute median area and time values for all nodes in software and hardware

while $|N_{free}| > 0$ do

For all nodes in N_{free}, compute M_i and t_i using GCLP with median area and time values.

Select a tagged node $T \in \{\text{ready nodes}\}$ using urgency measures

Use bin selection procedure to select the bin B_T^*

$N_{free} = N_{free} - T$, $N_{fixed} = N_{fixed} \cup \{T\}$

Update t_T based on selected implementation bin B_T^*

End.

9.6   Power Aware Partitioning on Reconfigurable Hardware

In this section, a hardware–software partitioning scheme has been presented that takes care of the overall power consumption of the system. The schedule generated is a time-valid and power-valid one, in the sense that the specified deadline for completion of the application(s) and the maximum power limit are honoured. A reconfigurable hardware (for example, FPGAs) can realize any functionality. Thus, if the target architecture consists of a general purpose processor capable of executing the tasks in software and a reconfigurable hardware-based platform that can again implement all the functionalities, from the power minimization point of view, the tasks be mapped to software for execution by the processor. However, this may lead to deadline violation. Thus, certain selected tasks should move to hardware. This, in turn, increases the power consumption of the system, leading to occasional power limit violation. Thus, careful choice has to be done about the nodes to be moved to the reconfigurable platform. Also, the software to hardware communication time and power needs to be considered in evaluating the overall system performance and power requirement. The architecture assumed here has been shown in Fig. 9.9.

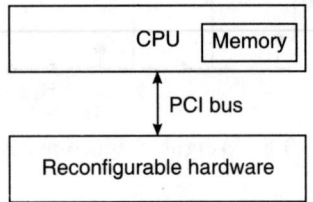

Fig. 9.9 Architecture for power aware partitioning.

The overall problem of power aware partitioning can be stated as follows:

Given an application set $\{A_1, A_2, \ldots, A_n\}$, *where each individual application* A_j *is specified as a directed acyclic graph with a periodic deadline* D_j, *total available hardware* Ah_{total} *and*

maximum power rating of the system P_{max}, generate a partition for each application so that they meet their timing and maximum power rating constraints with minimum hardware cost. More than one application could be active at one time.

The approach followed in this section is based on iterative improvement. It assumes that initially all tasks are mapped to software, and then iteratively moves tasks to hardware as long as the timing requirement is not satisfied. The power constraint is also checked in the process. We will first discuss about the partitioning of single functional systems—that is, systems with only a single application. The approach will be extended later for multiple applications.

9.6.1 Single Application Partitioning

The algorithm initially maps all the tasks in the application to software. The start times of the tasks on CPU are determined based on a priority function that ranks the tasks in the order of decreasing depth in the task graph. If the deadline is not satisfied, per iteration, one task is selected to be moved to hardware, based on the task's *mobility* indices and a *task selection routine*. We first present the task mobility.

The *mobility* of a task provides the information about parallelism that could be achieved by moving the task to hardware and thus achieving some level of parallelism. A task is said to be *mobile* if its latest possible start time (L_i) is more than the earliest possible start time (E_i). The earliest possible start time E_i is defined as,

$$E_i = max_{k \in pred(i)} \quad \eta(k)$$

where, *pred(i)* is the set of predecessors of task i and $\eta(k)$ denotes the finish time of task k. The latest possible start time L_i is similarly defined as,

$$L_i = min_{k \in succ(i)} \quad \{\eta(k) - .t_i\}$$

where, *succ(i)* is the set of successors of task i and $\eta(k)$ is the start time of task k. t_i is defined as follows,

$$t_i \quad = \quad ts_i \text{ if task } i \text{ mapped to software}$$
$$= \quad th_i, \text{ if task } i \text{ mapped to hardware}$$

The mobility of task i, $\mu(i)$ is defined as,

$$\mu(i) \quad = \quad 1, \text{ if } L_i > E_i$$
$$= \quad 0, \text{ if } L_i = E_i$$

Next, we present the task selection routine used at each iteration of the algorithm to select a task to move to hardware. In the routine, we assume that the variable N_s represents the set of software tasks in the application. The routine is as follows:

Procedure Task_Selection_Routine
Begin
    S.1 Rank the tasks in N_s in the order of decreasing software execution times ts_i
    S.2 Compute the mobility $\mu(i)$, $\forall i \in N_s$

S.3 If $\mu(i) = 0$ for all $i \in N_s$

Select task i with maximum execution time ts_i

else

Select task i with maximum execution time ts_i and non-zero mobility

End.

A schedule is said to be *power-valid* if the power profile of a single iteration of the application is less than or equal to the maximum power rating of the system. Otherwise, the schedule contains *power-spikes* which need to be removed from the schedule. Similarly, a schedule is said to be *time-valid* if all tasks in the application can be over within the deadline D for it. It may be noted that the communication delay also needs to be calculated. For the PCI-bus, it can be approximated as,

$$t_{comm} = \frac{\frac{CC * N_{sample}}{N_{bus}} + AC}{F}$$

where, CC is 2, the number of communication cycles needed per data element. N_{sample} is the total number of bits transmitted between the tasks, N_{bus} is the channel bit width, AC is the arbitration cycles needed, F is the frequency of the channel. The communication also leads to power dissipation in the bus, given by

$$P_{bus} = \frac{1}{2} \times C_{bus} \times V^2 \times m \times n$$

where, C_{bus} is the capacitance of the bus, V is the supply voltage, m is the number of words sent per second, n is the number of bits per word.

The communication between the tasks i mapped in software and j mapped in hardware (or vice versa) must be scheduled during the execution of task j, that is, between the earliest possible start time E_j and the latest possible finish time F_j of task j. It can be stated formally as,

$$\eta(i) + t_i \leq \eta(comm) + t_{comm} \;\; < \;\; F_j$$
$$E_j < \eta(j) + t_j \;\; \leq \;\; F_j$$

where, $\eta(i)$ and $\eta(j)$ are the start times of tasks i and j, t_i and t_j are the execution times of those tasks. E_j and F_j are the earliest possible start time and the latest possible finish time of task j. $\eta(comm)$ is the start time of communication and t_{comm} is the communication time between i and j, respectively.

The power-aware partitioning algorithm can be stated as follows:

Algorithm Power_Aware_Partitioning
Begin

Map all tasks to software. Schedule assumed to be power-valid

while schedule is not time-valid do

Select task i to be mapped to hardware using *Task_Selection_Routine*

Determine start time $\eta(i)$ using mobility $\mu(i)$

Update bus activity using the communication costs

Reschedule the task graph

Update total hardware used Ah

If (execution time decreases) and (schedule is power-valid) and ($Ah < Ah_{total}$) then

  Note new schedule; continue next iteration of loop

else if ($Ah > Ah_{total}$) then

  Invalidate selection of task i for hardware mapping for all future iterations

  Continue the loop

else if (schedule is not power-valid) then

  Invalidate selection of task i for hardware mapping for the next iteration

  Continue the loop

else if (no time improvement) then

  Invalidate selection of task i for hardware mapping for the next iteration

  Continue the loop

  end while

End.

9.6.2 Multiple Application Partitioning

The single application partitioning procedure can be extended to multiple applications by combining the tasks of all applications together into a *combined task graph* (CTG). Initially, all tasks are mapped into software. There may be similar tasks across the applications. Thus, selection of a task of one particular application to move to hardware may be influenced by similar tasks present in other applications. The order in which the applications are considered is based upon their *criticalities*. At each step of partitioning, the set of active applications $\{A_1, A_2, \ldots, A_n\}$ are ordered based on *criticality*. The criticality of an application (AC) is a measure of its computational complexity, described as,

$$AC_i = \frac{T_{CTG}}{D_i}$$

where, AC_i is the criticality of application A_i, D_i is its deadline and T_{CTG} is the completion time of one iteration of the combined task graph CTG.

The *task selection routine* is next modified to select a task to move to hardware from the CTG. It is driven by the hardware reusability, as similar tasks may be present in different applications. In this context, a *combined priority function* is defined for a task consisting of *self priority* and *shared priority*. The self priority is determined by the *mobility* of the task within the application, whereas, *shared priority* is based on hardware reusability of similar tasks in different applications. Thus, combined priority (CP) of a task i is the sum of its self priority, (SeP), and shared priority (ShP), defined as,

$$CP(i) = SeP(i) + ShP(i)$$

The computation of self priority for software tasks in an application can be performed as follows:

Procedure Self_Priority
Begin

    Compute mobility $\mu(i)$ for all $i \in N_s$, the set of software tasks of application A_k

    Determine $N_{s1} \subseteq N_s$, set of software tasks with nonzero mobility

    Determine $N_{s2} \subseteq N_s$, set of software tasks with zero mobility

    Initialize counter $Count = 0$

    while N_{s1} not empty do

        Extract task $i \in N_{s1}$ with maximum execution time t_{si}

        Compute self priority of task i as, $SeP(i) = \frac{N_s - Count}{N_s}$

        Increment $Count$

        Remove task i from N_{s1}

    end while

    while N_{s2} not empty do

        Extract task $i \in N_{s2}$ with maximum execution time t_{si}

        Compute self priority of task i as, $SeP(i) = \frac{N_s - Count}{N_s}$

        Increment $Count$

        Remove task i from N_{s2}

    end while

End.

The *shared priority* of a task i is computed as the total number of times a similar task across the set of applications has been mapped to hardware. If this number is Num_i, then *shared priority*, $ShP(i)$ is computed as the normalized quantity,

$$ShP(i) = \frac{Num_i}{max_{j \in N_s} Num_j}$$

Once the self and shared priorities have been computed, the combined priority can be computed as the sum of these two quantities. Based on this combined priority, the partitioning algorithm for multiple applications can be stated as follows:

Algorithm Multiple_Application_Partitioning

Input: Applications $A_1, A_2, \ldots, A_n$ along with deadlines $D_1, D_2, \ldots, D_n$

          Total available hardware area Ah_{total}

          Maximum power rating of the system P_{max}

Output: Time and power valid schedules for all applications

Begin

    while schedules for all applications are not time and power valid do

        Construct combined task graph CTG

        All tasks are initially mapped to software, schedule assumed to be power-valid

        For all $A_i \in \{A_1, A_2, \ldots, A_n\}$ compute *application criticality AC_i*

        Sort applications on descending order of criticality

        Select task i based upon *combined priority (CP)*

        Perform scheduling as in *Algorithm Power_Aware_Partitioning*

        Repeat for other applications in the ordered set of applications

For all $A_i \in \{A_1, A_2, \ldots, A_n\}$

      If schedule of A_i is both time and power valid then

            Remove A_i from the set

    end for

  end while

End.

9.7 Conclusion

In this chapter, we have seen different approaches to distribute the tasks of an application into the target platforms available—the general processors for software realization and FPGA/ASIC for hardware realization. The techniques are varied and based on different techniques, such as, integer linear programming, Kernighan-Lin partitioning, Genetic algorithms, Particle swarm optimization, etc. The GCLP algorithm has shown how to select between the available alternatives within a platform also. The power-aware partitioning scheme shows how to do the task distribution keeping in mind the overall power consumption of the system. In the next chapter we will look into the techniques to refine a task graph for an application to produce better implementation.

Exercises

9.1 Why is it important to partition a set of tasks into hardware and software platforms? What are the commonly used design criteria behind such a partitioning?

9.2 What is the main advantage of ILP-based approach over others for partitioning? What is its main drawback?

9.3 Explain the meaning of each of the resource constraints in the ILP-based partitioning.

9.4 Explain the meaning of each of the timing constraints in the ILP-based partitioning.

9.5 Justify the statement that "Kernighan-Lin is a hill-climbing heuristic".

9.6 Distinguish between a graph and a hypergraph. Study the Kernighan-Lin graph partitioning algorithm from some book on graph theory. Take an example graph and apply the algorithm on it.

9.7 Justify the statement that "Kernighan-Lin is an iterative improvement approach that produces balanced partitions". Why for hardware–software partitioning, the balanced partitioning requirement is not essential?

9.8 Which steps of Kernighan-Lin algorithm attempt to overcome the local minima? Illustrate the answer with suitable example.

9.9 Complete one more iteration of the example in Fig. 9.1.

9.10 Repeat Kernighan-Lin algorithm on Fig. 9.1 for area minimization.

9.11 Repeat Kernighan-Lin algorithm on Fig. 9.1 for area-performance joint minimization, in which equal weightage has been put on both area and performance.

9.12 Write down the change equations for node $n3$ of Fig. 9.1 for the movement of different nodes from software to hardware.

9.13 Write a C/C++ code for the Kernighan-Lin hardware–software partitioning. Show the partitioning results on randomly generated graphs. Generate other associated parameters, like size, execution time, parameters to be passed etc. also randomly.

9.14 How can genetic algorithm be used in hardware–software partitioning? Explain a possible structure of the chromosome for Fig. 9.1.

9.15 Write a C/C++ code to implement genetic algorithm based hardware–software partitioning. Compare the solutions produced with those for Exercise 9.13. Also compare their execution times.

9.16 Explain the concept of particle swarm optimization. How can it be applied in hardware–software partitioning?

9.17 Write a C/C++ code to implement particle swarm optimization based hardware–software partitioning. Compare the solutions produced with those for Exercises 9.13 and 9.15. Also compare their execution times.

9.18 What is extended partitioning? Why the schedule of tasks is also important for taking a partitioning decision?

9.19 Distinguish between binary partitioning and extended partitioning. What is meant by multiple implementation alternatives?

9.20 Study the list scheduling and explain it with an example.

9.21 Explain the concept of global criticality. How does it tune the partitioning process as more and more nodes are decided to be put in hardware or software?

9.22 Explain how the mapping objective is selected in GCLP algorithm.

9.23 What is meant by local phase? Why is it necessary to consider it alongwith global criticality? What are the different categories into which the nodes of a task graph can be partitioned. Give examples of each of these categories.

9.24 How are the extremity measures computed for a node? Classify the nodes in Fig. 9.1 assuming both α and β to be 50.

9.25 What are repeller measures? Mention some of the (i) software and (ii) hardware repeller properties. How to obtain the repeller measure of a node?

9.26 How is the problem of extended partitioning different from binary partitioning? What is meant by implementation bins?

9.27 What is bin fraction curve and how does it help in implementation bin selection?

9.28 Consider a set of adders implemented in different approaches, like ripple-carry, look-ahead etc. Implement the adders in some hardware platform and obtain the corresponding bin fraction curve.

9.29 Consider a Boolean function implementation in software. There can be several approaches—truth-table, sum-of-product, product-of-sum, Reed-Muller form, Binary Decision Diagram (BDD), etc. Take an example function with more than 20 inputs and represent it in these techniques. Then find the bin fraction curve.

9.30 What is bin sensitivity? Compute bin sensitivities for Exercises 9.28 and 9.29.

9.31 What is the utility of weighted bin sensitivity? Explain with an example. For Exercises 9.28 and 9.29 compute weighted bin sensitivities as well.

9.32 Why is power an important criteria for hardware–software partitioning? Should it be treated as a constraint or as an optimization criteria for partitioning?

9.33 Define mobility in the context of power-aware partitioning. Why in the Task_Selection_ Routine, if the mobility of all nodes are zero, a node with maximum execution time is selected to be moved to hardware?

9.34 Define the combined priority, self priority and shared priority.

CHAPTER 10

Functional Partitioning and Optimization

In the previous chapter, we have seen various strategies for distributing the tasks of an application among the available hardware and software platforms. This has essentially resulted in various schemes for hardware–software partitioning to optimize different design goals—area, delay, power, etc. However, at the specification stage of a system, the main emphasis is on a clean description of the system, understandable by the system designers. This helps in efficient management of the design in terms of its testing, validation, upgradation, modifications, and so on. The issue of a good implementability of the system is not a major concern. As a result, it is very much possible that the initial specification is not amenable to a good design, and necessitates several refinements to lead to another specification that results in good implementation. For example, the initial specification may be given in terms of a set of user-defined procedures. These procedures identify the conceptual module that the user visualizes the system to be consisting of. There may be commonality between the procedures, the procedures may be quite large or very small. Thus, taking each procedure as a task in the task-graph and trying to perform hardware–software partitioning over the set may lead to poor solutions. Though this initial set of procedures may act as a good starting point, the set needs to be regrouped and reconsidered for various optimizations possible. In this chapter, we will first look into the problem of functional partitioning of a large behavioural process into a set of procedures. Next, we will look into a set of optimizations that may be carried out over a behavioural specification to lead to better implementation.

10.1 Functional Partitioning

The process of functional partitioning divides a given functional specification of a system into a set of sub-specifications. Each sub-specification corresponds to a task that is ultimately mapped onto a custom hardware or to a software module to execute on a general purpose processor. Thus, it divides a large functional specification into a set of smaller ones. The approach has got several advantages as enumerated below:

1. It often results in order of magnitude reduction in the runtimes for logic synthesis. This happens because most of the heuristics used in logic synthesis work well for small-to medium-sized inputs. For large inputs, the quality of the solution degrades, also the runtime increases disproportionately. The tools are also not linear in runtime, in the

sense that, sum of runtimes for synthesizing several small processes can be much less than the runtime for one large process.

2. The overall system performance may also improve, if the smaller processes can be synthesized into custom hardware with small clock periods than the large hardware needed for the large specification, and when the shorter periods outweigh the overhead of inter-processor communication.

3. The partitioning also brings the scope of further parallelization between the tasks.

4. This may result in improved satisfaction of input/output and size capacity constraints on a package, such as FPGA. It can significantly reduce the number of inter-package signals, and may lead to fewer packages, smaller board size, reduced overall cost, etc.

10.1.1 Model

The input to the functional partitioning process is a single behavioural process X, which may, for example, be a C or VHDL code. The process describes the behaviour of a complex system, and the description consists of numerous operations, requiring many hundreds or thousands of lines of sequential program code. The process X can be viewed to be consisting of a set of procedures $F = \{f_1, f_2, \ldots, f_n\}$, with one being the main procedure which in turn calls other procedures. Any procedure can call others excepting the main procedure. A variable in the program is also treated as a simple procedure, with reads and writes being the procedure calls.

Functional partitioning of F creates a partition P consisting a set of parts $\{p_1, p_2, \ldots, p_m\}$, such that every procedure f_i is assigned to a single part p_j, that is, $p_1 \cup p_2 \cup \ldots \cup p_m = F$ and $p_i \cap p_j = \phi$ for all $i \in \{1, 2, \ldots n\}$ and $j \in \{1, 2, \ldots m\}$, $i \neq j$. Each part p_j will be implemented either by a custom hardware or as a software process running on a general purpose processor.

The model used can be the *call graph* of the process. Figure 10.1 shows an example model for a process *Mwt* consisting of a global variable *level* of type *byte*. It also has a number of other procedures as noted next.

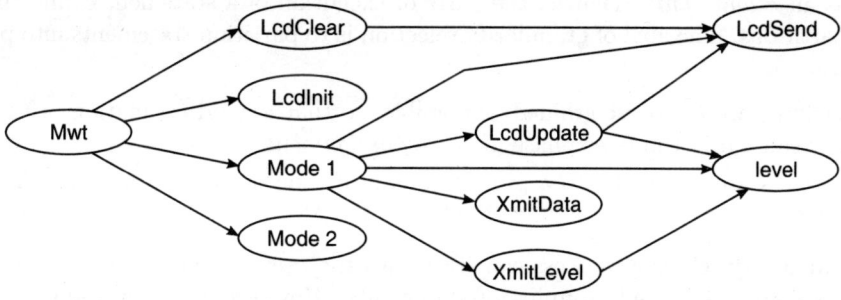

Fig. 10.1 Example model call graph.

Mwt
    byte *level*;
    — Procedures —
    Mode1()
    Mode2()

LcdInit()
LcdClear()
LcdUpdate(byte, byte)
XmitLevel(byte)
XmitData(bit)

10.1.2   Partitioning Methodology

Partitioning is a three-step procedure consisting of *granularity selection, preclustering,* and *N-way assignment* to modify a given access graph into a partitioned one. In the following paragraphs we discuss about each of them.

Granularity selection

Granularity is a measure of complexity of each procedure. *Fine granularity* means many procedures of low complexity, whereas *coarse granularity* means few procedures of high complexity. Selection of proper granularity can affect the partitioning process in the following ways:

1. A too fine granularity means creating a large number of procedures, which may make the later partitioning stage complex. It becomes difficult to estimate the performance of the partitioned system.

2. Creation of large number of procedures prevents the use of powerful partitioning heuristics that often have high complexities.

3. On the other hand, coarse granularity means that many behaviours are grouped together into a single procedure—an inseparable unit. Thus, it excludes many potential solutions that put those behaviours into separate groups.

Thus, the step of granularity selection must be carried out carefully. Each statement is taken as an *atomic unit.* That is, the parts of execution of a statement cannot be divided between partitions. The goal of granularity selection is to partition statements into procedures such that,

- procedures are as coarse-grained as possible, so that a good estimation of their performance can aid in the later phase of *N-way assignment.*

- statements are grouped into a procedure only if their separation would lead to inferior solutions.

The granularity selection process can start with the user-defined procedures, as they represent a good coarse-grained grouping of statements. However, these procedures need some improvements that can be achieved via transformations. The following are some such transformations:

1. *Procedure inlining:* This is a well-known transformation to replace a procedure call by the content of the called procedure with proper substitution of parameters. The inlined procedure disappears, while all calling procedures are made more complex. Eliminating the unnecessary procedure calls may keep the code that is executed close together in memory, improving the cache performance via improved locality of reference. This makes the granularity coarser.

2. *Procedure cloning:* It makes a copy of the called procedure for exclusive use by the procedure that called it. This is a compromise between inlining and not inlining. While inlining may lead to excessive size growth for multiply called procedure, keeping the called procedure in a separate part from the calling one may lead to communication bottleneck. Cloning will make a local copy of the called procedure which will eventually be put in the same partition as the caller by the later partitioning stage.

3. *Procedure exlining:* As the name suggests, it is the reverse of *inlining*. It essentially replaces a sequence of statements within a procedure by a call to a new procedure that contains only that subsequence. This makes the granularity finer. There are two types of exlining that can be tried out:

   - *Redundancy exlining:* It tries to replace two or more near identical statement sequences by a single procedure. For this purpose, the entire program is encoded as a string of characters. Now, an approximate string matching algorithm is used to identify the near identical substrings.
   - *Distinct-computation exlining:* It attempts to divide a large sequence of statements into several smaller procedures such that the statements within a procedure are tightly related. The statements are taken as atomic. A dependency analysis between the values computed by different statements may reveal the relationship between the statements. The process thus identifies a group of computations so that the statements within a group are highly related, whereas between the groups, the dependency is quite small.

Figure 10.2 shows an exlining of procedure *Mode1* into a new procedure *Mode1a*. The original procedure *Mode1* is complex in the sense that it handles two operations—LCD updation and transmitting data. It is divided into two procedures, the old procedure *Mode1* handles the LCD, whereas a call is made to the new process *Mode1a* that handles the data transmission operations. Assuming that a detailed analysis of the behaviour finds that the procedure *LcdClear* is called only once, which just calls *LcdSend* in turn, with a small number of parameters, the procedure *LcdClear* may be inlined with *Mwt*. This is also shown in Fig. 10.2.

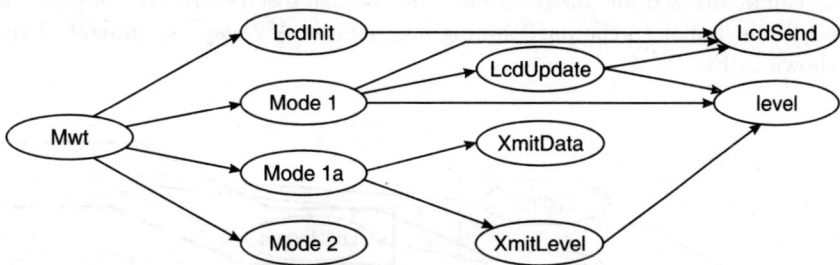

Fig. 10.2 Inlining of *LcdClear* and exlining of *Mode 1*.

Pre-clustering

The objective of this step is to reduce the number of procedures for *N-way partitioning*, by merging procedures whose separation among parts would never result in good solution. Unlike granularity selection, the procedures clustered here may not be such that they could have been inlined into a single procedure. On the other hand, it is different from *N-way partitioning* in the sense that each cluster does not represent a partition. A partition may include several clusters to be put into a part for allocation to a processor.

Pre-clustering is performed by computing a *closeness* value between the pairs of procedures resulting from the granularity selection. The closeness is defined as a weighted sum of several normalized closeness metrics, for example, the amount of parameter passing necessary, number of calls, etc. Figure 10.3 shows the results of pre-clustering. Assuming that the procedures *LcdUpdate* and *LcdSend* communicate very heavily, they should not be separated. On the other hand, inlining may not be a good option if *LcdSend* is called a large number of times. Thus, these two procedures are merged during pre-clustering, so that they are treated as one object subsequently.

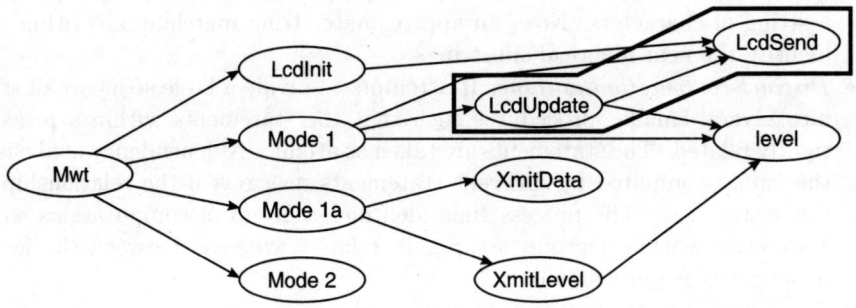

Fig. 10.3 Pre-clustering.

N-way assignment

The objective of this step is to distribute the procedures among the given set of processors—custom or general purpose. Many of the heuristics that we have seen for hardware–software partitioning can be utilized for this purpose. The detailed discussions are not given here, as the techniques will be similar to the partitioning algorithms. A 2-way assignment of the example has been shown in Fig. 10.4.

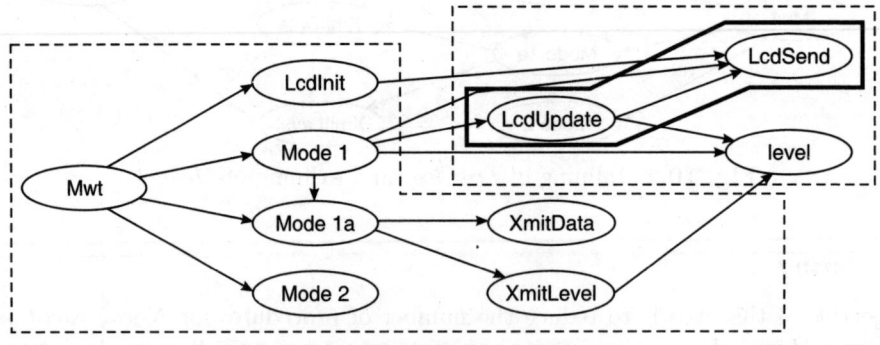

Fig. 10.4 2-way assignment.

10.2 High-level Optimizations

In this section, we will discuss about the optimization techniques that can be applied to a process description at higher level. We will mainly look into three powerful optimization techniques used in enhancing embedded specifications. These are:

- Loop optimizations
- Float to fixed-point conversion
- Data structure, such as array optimizations

10.2.1 Loop Optimizations

Among the various parts in a specification, loop optimization plays a vital role in improving the overall system performance. This is mainly because of the fact that the loop bodies are executed repeatedly within the operation of an embedded system. Thus, a small saving in terms of delay or power in the execution of a single iteration of a loop can be very much effective in producing a good saving of the quantity. There have been several strategies proposed for loop optimization. The commonly used techniques are the following:

- Loop interchange/permutation
- Loop splitting/peeling
- Loop fusion/combining
- Loop fission/distribution
- Loop unrolling
- Loop unswitching
- Loop-invariant code motion
- Loop tiling/blocking

In the following, we will discuss briefly about each of these techniques.

Loop interchange/permutation

It is the process of exchanging the order of two loops. One major utility of such a transformation is to improve the cache performance for accessing array elements. For example, consider the following code sequence:

$$\text{for } i = 1 \text{ to } 1000 \text{ do}$$
$$\text{for } j = 1 \text{ to } 1000 \text{ do}$$
$$a[i,j] = b[i,j] + c[i,j]$$

In this case, the array elements are accessed in the order of indices $(1,1)$, $(1,2)$, $(1,3)$, ... For a *row-major* organization (elements are stored row-wise in the memory) of the arrays, the memory locations are accessed successively. Thus, the code shows good performance improvement in the presence of caching of array elements. On the other hand, if the array elements are stored in a *column-major* form, the locality of reference is lost. Thus, it creates a large number of cache misses. To circumvent the problem, the loops may be interchanged in the following fashion, so that the elements are accessed as $(1,1)$, $(2,1)$, $(3,1)$, ...

$$\text{for } j = 1 \text{ to } 1000 \text{ do}$$
$$\text{for } i = 1 \text{ to } 1000 \text{ do}$$
$$a[i, j] = b[i, j] + c[i, j]$$

However, it is not mandatory that interchanging of loops will always lead to better cache utilization. For example, consider the following code fragment:

$$\text{for } i = 1 \text{ to } 1000 \text{ do}$$
$$\text{for } j = 1 \text{ to } 1000 \text{ do}$$
$$a[i] = a[i] + b[i, j] * c[i]$$

For a column major organization of the b-array, a loop interchange will bring locality into the accessing pattern of it, however, it will ruin the locality behaviour of a- and c-arrays. Also, it may not always be safe to exchange the iteration variables due to dependencies between the statements for the order in which they must execute.

Loop splitting/peeling

Very often, in the specification of a process, we have some condition checks nested deep within a set of loops. A typical example is in the field of image filtering. The filtered value of a pixel is often computed as a weighted sum of itself and its neighbours. Though for inner pixels, all the neighbours exist, for the boundary pixels, a number of neighbours may be absent. Thus, the typical processing runs over two loops as shown next.

$$\text{for } i = 1 \text{ to } num\_rows \text{ do}$$
$$\text{for } j = 1 \text{ to } num\_cols \text{ do}$$
$$\text{Check for boundaries}$$
$$\text{if boundary pixel then}$$
$$\text{do special processing}$$
$$\text{else do normal processing}$$

In this code, the check for boundary conditions is repeated for each non-boundary pixels as well. The loop can be split as follows:

$$\text{for } i = 2 \text{ to } num\_rows \, -1 \text{ do}$$
$$\text{for } j = 2 \text{ to } num\_cols \, -1 \text{ do}$$
$$\text{Normal processing}$$
$$\text{Special processing for all boundary pixels}$$

A useful special case of loop splitting is loop peeling that simplifies a loop with problematic first or few iterations separately before starting the loop. For example, consider the following code fragment:

$$p = 10$$
$$\text{for } i = 1 \text{ to } 10 \text{ do}$$
$$y[i] = x[i] + x[p]$$
$$p = i$$

Here, only for the first iteration p is equal to 10, for the rest it is $i - 1$. Thus, the loop can be modified as follows:

$$y[1] = x[1] + x[10]$$
$$\text{for } i = 2 \text{ to } 10 \text{ do}$$
$$y[i] = x[i] + x[i - 1]$$

Loop fusion and fission

Loop fusion/jamming is a transformation that replaces multiple loops by a single one. The reverse process is loop fission/distribution. Figure 10.5 shows the example of loop fusion and fission. Fission may be advantageous if the target processor provides zero-overhead loop instructions that can be used only in association with small loop bodies. On the other hand, loop fusion may lead to improved cache behaviour and possibility of parallelism within the loop body.

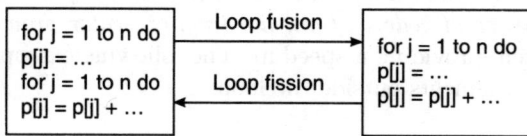

Fig. 10.5 Loop fission and fusion.

Loop unwinding/unrolling

This is a loop transformation technique that attempts to optimize the performance of a process at the expense of its size. The goal is to reduce speed by reducing (or eliminating) the *end-of-loop* test on each iteration. Loops can be rewritten as a sequence of statements and thus eliminate the loop controller overhead. The speed gained may offset the performance degradation due to the increased size of the code, and thus result in significant saving of processing time. Moreover, if the execution of the statements of the loop in an iteration are independent of previous iterations, all these statements can be executed in parallel. Thus, loop unrolling opens up the avenue of large-scale parallelism. The following example shows the case in which the loop has been unrolled once.

$$\text{for } j = 1 \text{ to } n \text{ do} \qquad \text{for } j = 1 \text{ to } n \text{ step 2 do}$$
$$p[j] = \ldots \qquad \Rightarrow \qquad p[j] = \ldots$$
$$p[j + 1] = \ldots$$

The number of copies of the loop is called the *unrolling factor*. For the above example, the unrolling factor is two. Higher unrolling factors are also possible. In the extreme case, loops can be completely unrolled, removing control overhead and branches altogether. Unrolling is normally restricted to loops with a constant number of iterations.

Loop unswitching

This essentially means moving a conditional from inside a loop body to the outside, by duplicating the loop body. A version of the loop body is placed inside each of the *if-* and *else-* clauses of the conditional. Since modern processors can operate fast on vectors, it can improve the overall speed of execution, though the code size is doubled, as the loop body is duplicated. The following shows the case of loop unswitching:

```
for i = 1 to 1000 do              if (w) then
    x[i] = x[i] + y[i]                for i = 1 to 1000 do
    if (w) then y[i] = 0                  x[i] = x[i] + y[i]
                          ⇒               y[i] = 0
                                  else
                                      for i = 1 to 1000 do
                                          x[i] = x[i] + y[i]
```

Loop-invariant code motion

Loop-invariant code refers to the statements which can be moved outside the body of the loop without affecting the semantics of the process description. The process of removing such code outside is called *loop-invariant code motion, hoisting,* or *scalar promotion*. The hoisted out code is executed less often, providing a speedup. The following example shows the removal of two loop-invariant code fragments outside the loop:

```
for i = 1 to n do                 x = y + z
    x = y + z                     t1 = x * x
    a[i] = 6 * i + x * x    ⇒     for i = 1 to n do
                                      a[i] = 6 * i + t1
```

Loop tiling

It is also known as *loop blocking, strip mine and interchange, unroll and jam,* and *supernode partitioning*. This optimization makes the execution of certain types of loops more efficient, particularly from the cache utilization point of view. For example, for the routines that involve handling large arrays, the cache size may not be sufficient to hold enough data for a single iteration of loops. In such cases, the loop indices are divided into subranges. Each subrange is processed by the iteration of inner loops. The outer loop indices change in the terms of subranges. A typical example of this is the *matrix-vector multiplication* algorithm shown below:

```
for i = 1 to N do
    for j = 1 to N do
        c[i] = c[i] + a[i, j] * b[j]
```

After a 2×2 tiling of the loops, the structure becomes,

```
for i = 1 to N step 2
    for j = 1 to N step 2
        for ii = i to min(i + 2, N)
            for jj = j to min(j + 2, N)
                c[ii] = c[ii] + a[ii, jj] * b[jj]
```

Figure 10.6 shows a $B \times B$ tiling for the full matrix multiplication.

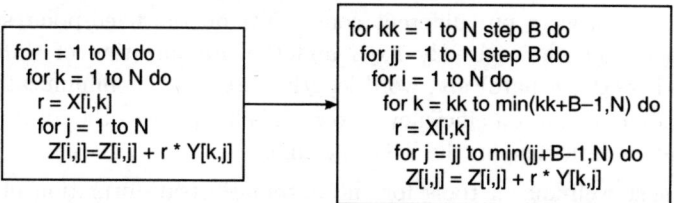

Fig. 10.6 Loop tiling.

10.2.2 Floating to Fixed Point Conversion

For most digital systems, the design must result into a fixed-point implementation, as fixed-point realizations require significantly less power, chip size and price per device, compared to floating-point ones. A typical system design flow has been shown in Fig. 10.7. The major obstacles in the conversion process are the following:

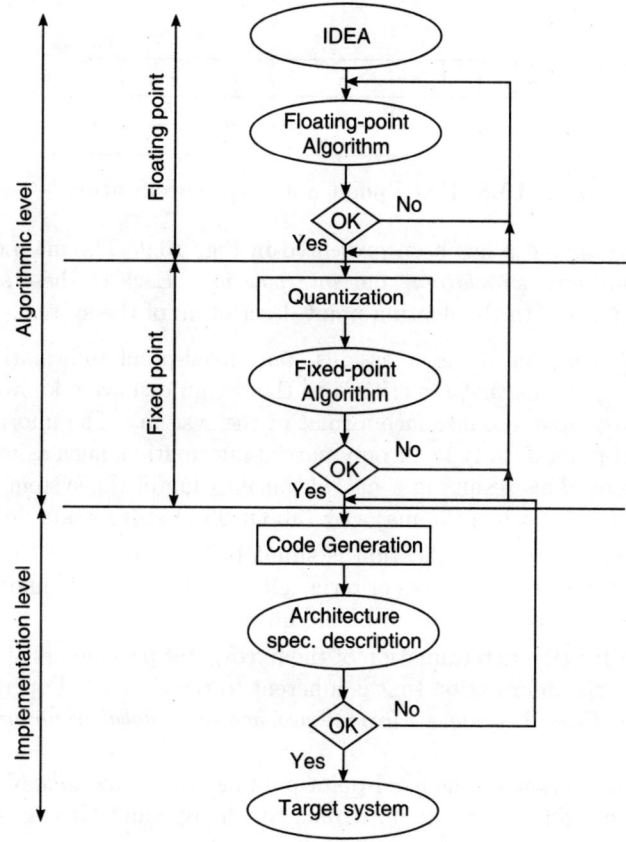

Fig. 10.7 Fixed-point design process.

1. The transformation from floating-point algorithm to fixed-point algorithm is often manual, which is time-consuming and error-prone.

2. Hardware and software put different constraints on the fixed-point specification. For software, word length is already fixed and the minimization of shift operation is of interest. Whereas, for hardware, word length is free and its minimization is a concern.

3. The fixed-point simulation efficiency is low. This happens as the fixed-point is not available as built-in data type on the host machine.

One of the most well-known tools for the automatic transformation of a program with floating-point specification into one with fixed-point specification is FRIDGE (*F*ixed-point p*R*ogramm*I*ng *D*esign *E*nvironment). It is based on an *interpolative approach* discussed next.

Interpolative approach

A fixed-point specification of an algorithm requires the assignment of a three-tuple to each of its operands, as shown in Fig. 10.8:
- *wl*—the assigned word-length
- *iwl*—the number of bits alloted for the integer portion
- *s*—the sign bit.

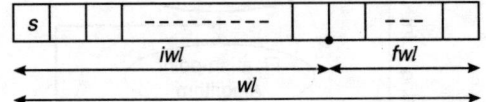

Fig. 10.8 Fixed-point data type specification.

The *interpolative approach* has been presented in Fig. 10.9. The main components are as follows—*local annotations, simulation,* and *interpolation.* Each of these steps are discussed next. The process starts with the floating-point description of the system.

1. *Local annotations:* The designer assigns some fixed-point information to some of the floating-point operands that are critical to the design or have a known fixed-point specification (for example, the interface format of the system). The information may be the complete fixed-point data type, or only partial information, such as word length, integer word length, etc. This results in a *hybrid* specification of the system consisting of some fixed-point operands, while the majority (about 95%) still remain floating point.

2. *Simulation:* The hybrid specification is simulated to check that the locally annotated specification still meets the design criteria. Otherwise, the local annotations, or even the structure of the algorithm needs to be modified.

3. *Interpolation:* It is the determination of the fixed-point parameters of the non-annotated operands from the information that is inherent to the annotated operands. The concept is based upon three key ideas—*format propagation, global annotations,* and *designer support.*

   (a) *Format propagation:* The fixed-point parameters *iwl* and *sign* of an operand can be determined from the range [*min, max*] of the operand. It is given by the relationships,

$$sign = \begin{cases} u & \text{if } min > 0 \\ s & \text{otherwise} \end{cases}$$

$$iwl = \max\{\lceil \log(min) + 1 \rceil, \lfloor \log(max) \rfloor + 2\}$$

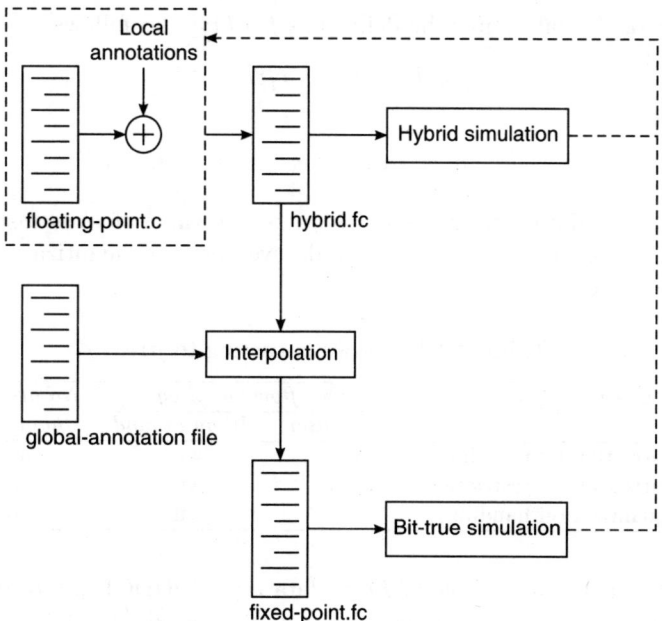

Fig. 10.9 Interpolative approach.

One extra bit is kept in *iwl* to guarantee the representation of the maximum and the minimum values. The value of *fwl* can be determined to represent the necessary fractional bits that carry any information. From this, the word length *wl* can be determined. Given the information about the operator and the fixed-point format of the inputs of operation, the range and fractional word length of the result can be determined. Format propagation requires an analysis of the data- and control-flow of the program at compile time to gather the fixed-point parameters.

(b) *Global annotations:* These are the restrictions to be matched throughout the complete design, irrespective of individual operators and operands. For example, a 16×16 multiplication produces 32-bit result. At no point of time should the word length exceed 32 bits. For ASIC implmentation, this restriction will be applied throughout and whenever the result exceeds 32 bits, it may be set to a default size of 32 bits.

(c) *Designer support:* At some places of code, interpolation may not be possible to be carried out automatically. In such cases, the system designer may be asked to provide some necessary values for the interpolation to continue.

4. *Simulation:* The global annotations might have changed the algorithmic performance of the specification. Thus, the fixed-point program needs to be simulated again. If the system does not fulfill the design criteria, the initial description might be modified by adding or changing annotations.

The tool FRIDGE converts a floating-point program written in ANSI-C to one with fixed-point data types. It supports two new data types—*fixed* and *Fixed*.

1. *Data type fixed:* Variables may be defined as *fixed* type as follows.

$$fixed \qquad a, \; {}^*b, \; c[8]$$

A statement like,

$$a = fixed(wl, \; iwl, \; sign, \; cast, \; {}^*b)$$

means, a receives data type $< wl, iwl, sign >$, the value of $*b$ is casted based on *cast*. The casting mode identifies how to handle overflow and quantization. This has been shown in Table 10.1.

Table 10.1 Modes of casting to *fixed*

Casting mode	Overflow handling		No overflow handling
	Saturation	Wrap around	
quantization by rounding	sr	wr	nr
quantization by restriction	st	wt	nt
no quantization handling	sn	wn	nn

2. *Data type Fixed:* For a variable of *Fixed* data type, FRIDGE performs data type check for every assignment to the variable. For example, consider the following code fragment:

$$Fixed{<}6,3,s{>} \; d;$$
$$d = fixed(7,4,s,sr,{}^*e);$$

The right hand side of the assignment has the format $< 7, 4, s >$, which is incompatible with the type of d declared previously. FRIDGE may give a warning, or force execution interrupts depending upon the mode set.

The interpolation carried out by FRIDGE is similar to that discussed earlier. Some salient features are noted next.

1. *Sequential code:* For constants in a sequential code, the minimum width requirements for wl, iwl and sign s are determined without losing any information. For variables, by interpolation, it tries to determine the word length that is sufficient. For example, consider the following piece of code given as input to FRIDGE.

```
global_cast(sr);
float b = 2.75;
float c = 5.0;
float a;
a = c - b;
```

The output produced by FRIDGE is as follows:

```
fixed b = fixed(4, 2, u, sr, 2.75);
fixed c = fixed(3, 3, u, sr, 5.0);
fixed a;
a = fixed(4, 2, u, sr, c - b);
```

2. *Conditional structures:* Depending upon the execution, a variable may be instantiated differently. For example, consider the following input segment given to FRIDGE.

global_cast(sr);
float $b = 2.75$;
float $c = 5.0$;
float a;
if (condition)
    $a = c - b$;
else
    $a = 1.875$;
$d = a$;

The output produced by FRIDGE is as follows:

fixed $b = $ fixed(4, 2, u, sr, 2.75);
fixed $c = $ fixed(3, 3, u, sr, 5.0);
fixed a;
if (condition)
    $a = $ fixed(4, 2, u, sr, $c - b$);
else
    $a = $ fixed(4, 1, u, sr, 1.875);
$d = $ fixed(5, 2, u, sr, a);

3. *Loop constructs:* FRIDGE analyses the number of iterations. For each iteration, it determines the necessary fixed-point parameters separately. If it identifies that the parameters are not equal for all iterations, it automatically generates an array containing the iteration specific fixed-point format. These arrays are accessed via pointers. The following is an example of the same:

global_cast(sr);
float $b[2] = 2.75, -3.5$;
float a;
$a = 0$;
for $(i = 0; i < 2; i + +)$
    $a = a + b[i]$;

The output produced by FRIDGE for this input is as follows:

fixed $b[2] = \{$fixed(4, 2, u, sr, 2.75),
                fixed(4, 3, s, sr, -3.5)$\}$;
fixed $c = $ fixed(3, 3, u, sr, 5.0);
fixed a;
int $w1\_a[] = 5, 3$;
int $*pw1\_a = w1\_a$;
int $iw1\_a[] = 3, 1$;
int $*piw1\_a = iw1\_a$;
$a = 0$;
for $(i = 0; i < 2; i + +)$
    $a = $ fixed($*pw1\_a++, *piw1\_a++,$ s, sr, $a + b[i]$);

Thus, the interpolator is a very powerful one based on strong data flow and control flow analysis. It also covers pointers, arrays, and static variables.

10.3   Conclusion

In this chapter, we have seen how to refine a specification to make it more amenable towards better realization. Two major techniques have been discussed. The first one refines a set of procedures by breaking and regrouping them keeping in consideration their closeness. Next, we have looked into a set of tranformations that improves the performance of the system. We have also seen a tool that can convert a floating-point algorithm into a fixed-point one, without losing accuracy. In the next chapter, we will look into another very important aspect of embedded systems—the power consumption.

Exercises

10.1 What is meant by functional partitioning of a specification? How is it expected to affect the hardware–software partitioning?

10.2 Mention the three steps of functional partitioning and discuss about their responsibilities.

10.3 How does granularity selection affect the partitioning process?

10.4 Explain procedure inlining, cloning and exlining with examples.

10.5 What is the difference between procedure inlining and clustering?

10.6 What is meant by loop permutation? Assume that the cache can hold 10000 integers at a time and that in case of a cache miss, it loads 10000 integers. For the following program fragments, identify the number of cache misses in a row-major organization of arrays:

    (a)   for $i = 1$ to 20000 do
                for $j = 1$ to 20000 do
                      $a[i, j] = a[i, j] + i * j$

    (b)   for $j = 1$ to 20000 do
                for $i = 1$ to 20000 do
                      $a[i, j] = a[i, j] + i * j$

10.7 What is meant by loop splitting/peeling? For the following loop, perform loop peeling.

        for $i = 1$ to 1000 do
            for $j = 1$ to 1000 do
                if $i = 1$ then $a[i, j] = 0$
                else if $j = 1000$ then $a[i, j] = 1$
                else $a[i, j] = i * j + 5$

10.8 Explain loop fusion and fission. In which type of target processor these are going to be advantageous? With the same assumptions about cache as in Exercise 10.6, determine the number of cache misses in the following two cases:

(a) $Sum = 0$
        for $i = 1$ to 20000 do
            for $j = 1$ to 20000 do
                $Sum = Sum + a[i, j]$
(b) $Mult = 1$
        for $j = 1$ to 20000 do
            for $i = 1$ to 20000 do
                $Mult = Mult * a[i, j]$

10.9 What is loop unrolling? What is meant by unrolling factor? Why is it the case that unrolling is generally restricted to loops with fixed number of iterations?

10.10 What is loop-invariant code motion? How does it help in improving the performance of a system? For the following code fragment, identify the invariant portions and move them outside the loop(s).

    for $i = 1$ to 1000 do
        $k = 100$
        $p = i * i$
        for $j = 1$ to 1000 do
            $m = p + i$
            $x = 10$
            $a[i, j] = a[i, j] + m * k - p * x$

10.11 What is loop tiling? How does it help in improving the system performance?

10.12 Why is it desirable to convert floating-point quantities into fixed-point ones? What are the major difficulties in the process?

10.13 What is the interpolative approach in fixed-point conversion process? Enumerate the steps involved in the process.

CHAPTER 11

Low Power Embedded System Design

Power consumption is a very important issue in embedded system design. Many of the embedded applications are built around batteries, and thus, battery life is a critical concern in judging their acceptance. Some of the important issues to consider include power consumption limits, size restrictions, I/O requirements, operational duty cycle, etc. There are several practical considerations that are involved in the process.

1. *Recharging facility:* It is not always possible to recharge/replace the batteries of embedded systems. This is particularly true for systems employed at remote and difficult-to-reach places, for example, a data collection centre for wildlife.

2. *Device size:* Unfortunately, battery technology has not scaled down at the same rate as IC technology. Thus, the weight and size of the device is often guided by the required battery capacity and its associated size. This is a very practical problem for mobile/portable devices.

3. *Duration of operation:* The duration for which a device needs to be active, is another major issue determining the total energy consumption. If the device is idle for majority of the time, a power-down mode may be incorporated into the design to save power.

4. *Power required by the device:* This is the most important issue and needs to be estimated even at the stage of initial design. A better estimation helps in the design process to try out alternatives.

5. *I/O device types:* The I/O interfaces, such as optically isolated I/O and electromechanical relays consume high power. Thus, the system designer needs to avoid them altogether, or minimize their active periods.

6. *Speed of operation:* As we will see later, power consumption is directly proportional to the frequency of operation. The system designer needs to identify the optimum speed for each component, so that the performance target is met, at the same time, individual subsystems are not operated at a speed higher than the required one.

11.1 Sources of Power Dissipation

Since majority of the systems designed are built around CMOS, in this section we will look into the different sources of power dissipation in CMOS. The total power consumed can broadly be divided into *dynamic power* and *static power*.

11.1.1 Dynamic Power Dissipation

This refers to the power consumed by a circuit/system due to some activities within it. For a *static CMOS* realization, at steady-state, either the pull-up (p) or the pull-down (n) network is OFF. Thus, when there are no activities in the system, power consumption is expected to be very low. During circuit activities, the power consumed depend upon the following two mechanisms.

1. *Short-circuit power:* If there is a finite rise and fall time at the input of a CMOS gate, both p- and n-networks are ON simultaneously, shorting the power supply line to the ground. If V_{DD} is the supply voltage and I_{mean} is the average current drawn during the input transition, then the short-circuit power is given by,

$$P_{shortcircuit} = I_{mean} \times V_{DD}$$

   For properly sized and ratioed gates, the contribution to the overall dynamic power due to $P_{shortcircuit}$ is of the order of 10–20%.

2. *Switching power dissipation:* This is the power consumed due to charging and discharging of capacitive loads when the circuit has some activities due to change in inputs. The capacitive load at different circuit gates depends upon the fanout of the gate, output capacitance, and wiring capacitances. It may be noted that a node with load capacitance might not switch when the clock is switching. To take care of this, a quantity called *switching activity* (α) is often used. It determines how often switching occurs on a node with load capacitance. If V_{DD} is the supply voltage, V_{swing} is the change in voltage level of the switched capacitance, C is the capacitance being switched and f is the frequency of operation, the switching power is given by,

$$P_{switching} = C \times V_{DD} \times V_{swing} \times \alpha \times f$$

   Since in most of the cases, $V_{swing} = V_{DD}$,

$$P_{switching} = C \times V_{DD}^2 \times \alpha \times f$$

11.1.2 Static Power Dissipation

This is the power consumed when the circuit is not in active mode of operation. In such a situation, there is still some power dissipation due to various leakage mechanisms. The situation is aggravated with the scaling of supply voltages. As the supply voltage is reduced, to keep the delay of a gate unchanged, the transistors need to be turned ON early by reducing their threshold voltages. This, in turn, increases the leakage current in the subthreshold range of operation of the circuit. Due to the exponential nature of leakage current in the subthreshold regime of the transistor, it can no longer be ignored. The *International Technology Roadmap for Semiconductors* (ITRS) has projected an exponential increase in leakage power with minimization of devices. Also, leakage increases with temperature. Thus, the increased heat dissipation resulting from increase in leakage power consumption has a positive feedback on leakage. There are three major leakage mechanisms—*subthreshold leakage, gate direct tunneling* and *junction band-to-band tunneling*. This has been shown in Fig. 11.1.

- *Subthreshold leakage:* When gate voltage is below threshold voltage but very close to it, subthreshold conduction current flows between source and drain. It is caused by the diffusion of minority carriers. It depends exponentially on the threshold voltage of the transistor. In nano-scaled devices, *short channel effect* (SCE) reduces the threshold voltage, thereby increasing the subthreshold current.

- *Gate direct tunneling leakage:* Due to ultra-thin gate oxide, a high electric field can cause electrons to tunnel through the gate-oxide. This results in large gate leakage in nano-scale transistors. An increase in the supply voltage and/or reduction in oxide thickness results in an exponential increase in gate tunneling current.

- *Junction band-to-band tunneling leakage:* Application of reverse bias across the highly doped p-n junction results in tunneling of electrons from valence band of p-side to the conduction band of n-side. This is called *band-to-band tunneling*. In nano-scale devices, due to the use of high junction doping, large junction BTBT occurs at OFF state with the drain at V_{DD} and substrate at ground. The junction BTBT increases exponentially with an increase in junction doping and supply voltage.

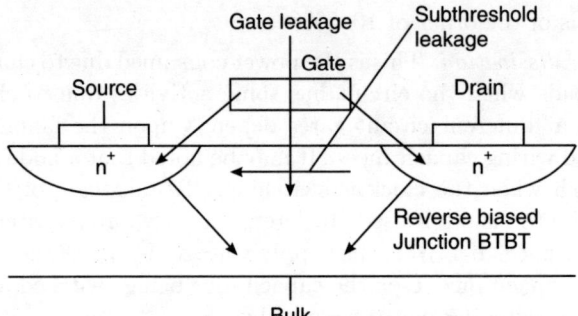

Fig. 11.1 Leakage components in a transistor.

11.2   Power Reduction Techniques

Power reduction can be attempted at all levels of design hierarchy—*algorithm, architecture, logic, and device* levels. In the following sub-sections we give an overview of each of these (except the device level on which the embedded system designer often does not have any control). Higher the level at which power minimization is addressed, higher is the expected power saving.

11.2.1   Algorithmic Power Minimization

It mainly focuses on reducing the number of operations requiring larger power in a target implementation. For example, in many processors, the cost of an addition/subtraction operation may be different from a logical operation. Thus, to check *"whether x is equal to y"*, one may first perform a subtraction operation followed by checking the status register for zero-bit. On the other hand, if the logical operation takes lesser power, x may be directly compared with y using a comparison instruction. The following are some of the important issues to be judged for selecting a particular algorithm from alternatives:

1. *Memory reference:* This is very important as memory is normally off-chip from the processor. A large number of accesses to the memory mean good amount of activity in

the address/data bus lines. The memory access pattern is also important. If the access pattern is sequential, only the least significant bits of address bus change, whereas for random access through the memory, most of the address bits will switch, thus creating higher power dissipation.

2. *Presence of cache memory:* The presence and structure of cache memory plays an important role. Cache can be fruitfully utilized to reduce both execution time and power of an implementation if the underlying algorithm has got *locality* in its behaviour. The locality may be both *temporal* and *spatial* in nature. While a temporal locality refers to the fact that a memory location accessed at some time is also likely to be accessed in near future, spatial locality means if a memory location is accessed at some time, its neighbouring locations are also likely to be accessed in near future. Thus, caching them inside the CPU cache saves not only the memory access time, but also the bus energy consumption is reduced.

3. *Recomputation vs. memory load/store:* Normal power minimization techniques at algorithm level attempt to reduce the number of arithmetic operations. However, it may so happen that to reduce the number of operations, some repeatedly performed computation is done only once and stored at a memory location. Later, as and when necessary, it is reloaded from the memory. This may lead to increased power consumption due to extra memory accesses. If the operands are already available in CPU registers or on-chip cache, it may be better to recompute the value, instead of loading it from memory, from power consumption point of view.

4. *Compiler optimization technique:* The typical techniques used by an optimizing compiler can be used to reduce power consumption of a piece of code. The strategies involve *strength reduction, common subexpression elimination, minimizing memory traffic* etc. *Loop unrolling* is also often beneficial as it reduces loop overhead.

5. *Number representation:* This is another area for algorithmic power trade-off. The following points may be noted:

   - *Fixed vs. floating point representation:* Fixed point operations are much simpler than floating point ones. Thus, it normally leads to power saving, though accuracy may suffer.
   - *Sign-magnitude vs. 2's complement:* Selection of sign-magnitude representation may have significant power saving over 2's complement, if input samples are uncorrelated and range is minimized.
   - *Precision of operations:* This is important, since having lower precision allows one to reduce the size of space needed to store the values. A typical example of this is to reduce the number of bits in mantissa portion in several signal processing applications including speech and image to improve circuit delay and power.

11.2.2 Architectural Power Minimization

The architectural level transformations can be used to introduce power saving in a design. There are two generic techniques to save power at the cost of extra area, keeping the performance of the system unaltered. These are *parallelism* and *pipelining*. To understand their impact, suppose we have to design a system with supply voltage V and operating frequency f. Figure 11.2(a) shows such a system at a black-box level. The system is expected to operate on a sequence of input data arriving at a rate of f. Now, if we duplicate to have n such similar modules and the input data is processed by the modules in an interleaved fashion, the blocks may operate at a lower speed—ideally, at a rate of f/n. Since the system is capable of running at frequency f and the speed of a system is proportional to the supply voltage, we reduce the supply from V to V/n. This will effectively reduce the power consumed by the individual blocks by a factor of n^2 (as power consumption is proportional to the square of supply voltage). Since all such systems are operating simultaneously, total power saving is $1/n$ of the original power. This has been shown in Fig. 11.2(b). A problem with the scheme is that the hardware is duplicated with other necessary multiplexing and demultiplexing logic. Another possible architectural modification often suggested is *pipelining*. In this scheme, the functional block of Fig. 11.2(a) is divided into a sequence of sub-blocks, each of approximately same delay. Thus, if the number of sub-blocks be n, from pipelining principle, the overall system can produce output at a rate of about $n \times f$. Now, if the supply voltage of individual stages is reduced by a factor of n, power reduces by a factor of $1/n^2$. However, we need to accommodate extra latches between the stages for proper synchronization between them. This introduces some overhead in terms of area, performance and power as well. The scheme has been shown in Fig. 11.2(c).

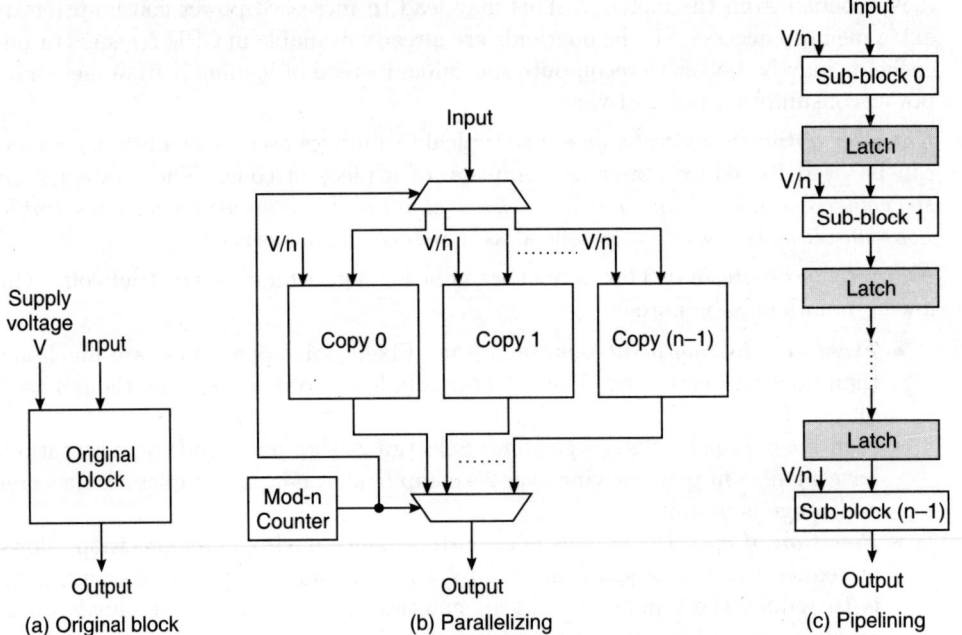

Fig. 11.2 Architectural transformations for low power.

11.2.3 Logic and Circuit Level Power Minimization

The power reduction approaches at logic and circuit level mostly target to reduce the effective switched capacitance. The following are some of the directions in which this minimization can be tried out:

1. *Static vs. dynamic logic families:* CMOS logic can be realized as static or dynamic CMOS. Depending upon the signal transition probabilities, one of the design styles may have an edge over the other. For example, for a two-input NAND gate with uniform probability distribution for the inputs, the probability that the output is zero is 0.25, whereas, the probability of output being one is 0.75. Thus, the probability of a power consuming $0 \to 1$ transition is $0.25 \times 0.75 = 0.1875$. On the other hand, for a dynamic circuit, output is always precharged to 1. Thus, power will be consumed whenever output is zero. Hence, the probability of a power consuming transition is 0.25, which is higher than a static gate. However, dynamic gate has lower input capacitance (almost by a factor of 2 to 3) compared to static gate, as the p-network is absent. Hence, the effective capacitance that a dynamic gate sees is much lower. But, the power consumed in distributing the precharging signal also needs to be considered.

2. *Glitches and hazards:* This is another potential source of power consumption, particularly in static CMOS circuits. A glitch at the output of a gate can come due to the differences in arrival times of input signals. A typical example of it is AND-OR-INVERT based implementation of the function $f = ab + \bar{a}c$. The circuit is shown in Fig. 11.3(a). Assume that b and c are always 1, while a is making a transition from 1 to 0. Ideally, output should remain fixed at 1, however due to the delay introduced by the inverter, the output z will continue to be 0, when y is also 0. This changes the output f to 0. However, after that small delay of the inverter, z again becomes 1, thus making $f = 1$. Thus, y, instead of remaining fixed at 1, will show a transition $1 \to 0 \to 1$, introducing a glitch. The glitch can be removed by adding a redundant AND gate for the term bc and feeding the output to the OR gate, as shown in Fig. 11.3(b). It may be noted that dynamic logic does not suffer from any glitch power consumption since all inputs must be valid before the gate evaluates.

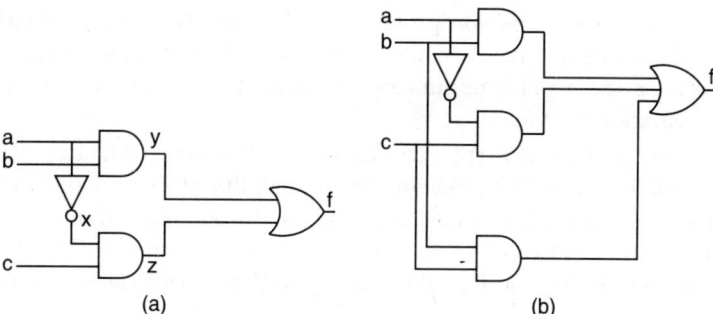

Fig. 11.3 (a) Circuit with hazard, (b) Circuit without hazard.

3. *Technology mapping:* The logic synthesis library often contains different implementations of the same logic module. They normally differ in terms of area, delay, power, etc. A logic synthesis procedure targeted to power minimization may choose implementations that require higher area or delay, but score better in terms of power. For example, consider a four-input AND function. Two possible implementations are shown in Fig. 11.4(a) and Fig. 11.4(b), respectively. The ON-probabilities of the gates are also shown. Total $0 \to 1$ transition probability of the first implementation is $0.25 \times 0.75 + 0.25 \times 0.75 + 0.9375 \times 0.0625 = 0.4336$. For the second implementation, it is, $0.25 \times 0.75 + 0.125 \times 0.875$

$+ 0.9375 \times 0.0625 = 0.3555$. Thus, the first implementation consumes more power than the second one. In this case, though there is no area penalty, the second implementation has one gate delay more than the first one.

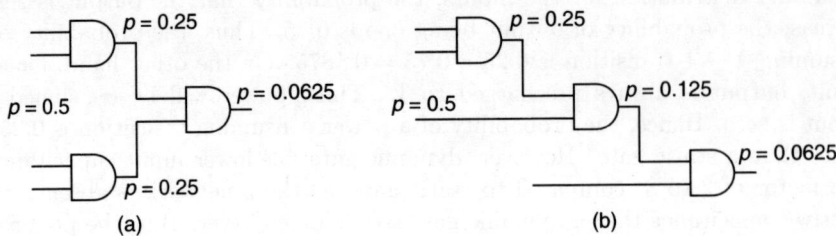

Fig. 11.4 Two different implementations of 4-input AND.

11.2.4   Control Logic Power Minimization

Power optimization of controller circuitry is very important. This is because while the datapath of a design can be selectively turned OFF when not in use, controller is always active. A controller is often specified and realized as a *Finite State Machine* (FSM). The synthesis of FSM targeting low power can be addressed from various angles as discussed next.

1. *Storage element design:* Several flip-flop design have been reported in the literature with various area-power trade-offs. Detailed discussion on this is beyond the scope of this book.

2. *State assignment:* This is the assignment of binary codes to the FSM states to realize it. For low power state encoding, first the steady-state probability for each of the states is determined. Next, the transition probabilities between the states are calculated. Codes with lesser Hamming distance are allocated to the states having higher transition probabilities between them. This minimizes the number of transitions in the next-state and output combinational logic.

3. *FSM partitioning:* Often it is the case that the FSM states form clusters. Once the FSM is in some state belonging to a cluster, the probability of its remaining within the states of this cluster for quite a few transitions is high. Probabilities of inter-cluster transitions are low. Thus, the FSM can be partitioned into two or more sub-FSMs. At any point of time, only one sub-FSM is active. We can do two things with the remaining submachines. The first alternative is to stop clocks to them and make sure that their primary input lines are masked-off. This is commonly known as *clock gating*. The scheme has been shown in Fig. 11.5. Thus, there is no dynamic power consumption in these sub-FSMs. The other alternative is to use *power gating* by introducing *sleep transistors* to turn OFF power supply to them. Figure 11.6 shows introduction of sleep transistors. Power gating reduces leakage power as well, however, wake-up takes time. Thus, either the circuit performance suffers, or we have to turn ON probable active sub-FSMs early—this leads to wastage of power as more than one sub-FSMs are active.

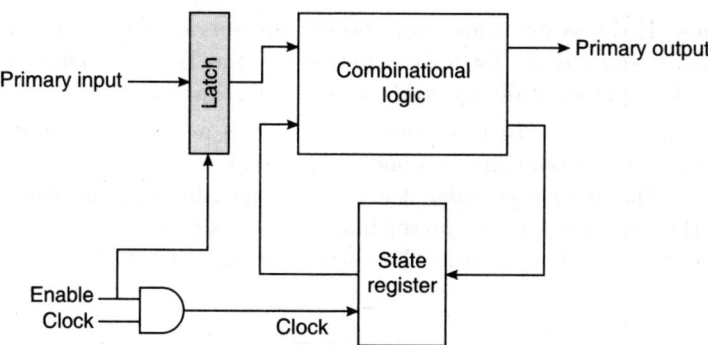

Fig. 11.5 Clock gating of FSM.

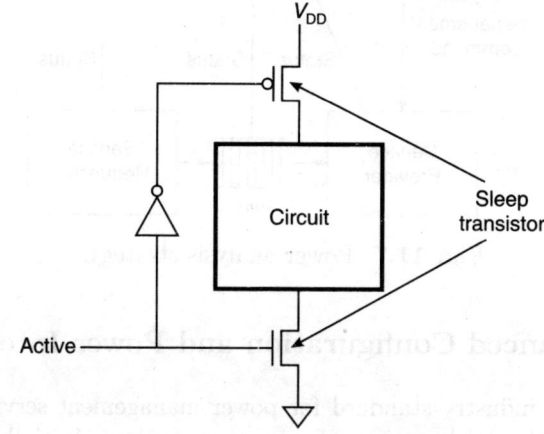

Fig. 11.6 Power gating of FSM.

11.3 System Level Power Management

The techniques discussed so far are good to the extent that they deal with local subsystems. In an embedded system, consisting of various subsystems, it may be possible that all of them are not needed simultaneously to remain active. Thus, it needs a system level management policy to turn OFF and turn ON the subsystems. A suitable power management policy is needed for the following purposes:

- going to a low power state takes time. The longer the duration for which we want to shutdown a system, higher is the time taken during reactivation.
- avoiding a power-down mode will cost unnecessary power.
- frequent power-down mode will affect system performance.

A naive approach may be to power-down a system whenever there is no request. This will definitely affect performance severely. A more sophisticated method is to use *predictive shutdown*. In this approach, the goal is to predict the next arrival of service request and wake up the system just before that. Prediction can be made in several different ways as follows.

1. *Fixed times:* If the system does not receive any service request during an interval of length T_{ON}, it shuts down for a fixed period of time T_{OFF}. Choice of T_{ON} and T_{OFF} may be made experimentally by studying system behaviour.

2. *Analysing system state:* In this approach, there is a constant monitoring of the service requests. The monitoring is done via a *power manager* that observes the system components—the *service provider*, the *service requestor* and the *queue* between them. Based on the observations, the power manager sends power management commands to the service provider. The situation has been shown in Fig. 11.7.

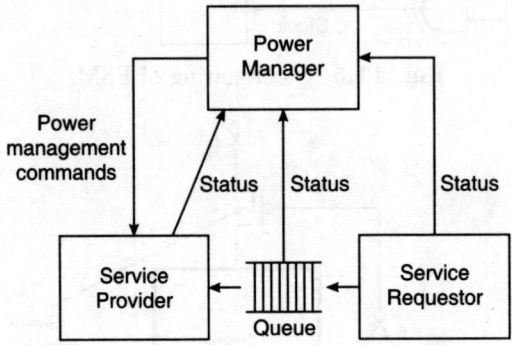

Fig. 11.7 Power analysis strategy.

11.3.1 Advanced Configuration and Power Interface (ACPI)

ACPI is an open industry standard for power management services. It is designed to be compatible with a wide variety of operating systems. Initially targeted at *Personal Computers*, ACPI provides some basic power management facilities and abstracts the hardware layer. Host operating system will have its own power management module that determines the policy. The operating system utilizes ACPI to send the required controls to the ACPI compliant hardware and observe the hardware's state as input to the power manager. The concept has been shown in Fig. 11.8.

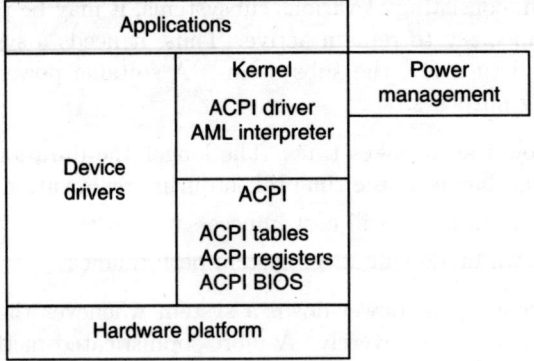

Fig. 11.8 ACPI interface.

ACPI supports five basic global power states, as follows:

- *G3:* The mechanical OFF state, in which system consumes no power.
- *G2:* The soft OFF state, which requires a full operating system reboot to restore the machine to working condition. It has four substates:
    - *S1:* a low wake-up latency state with no loss of system context.
    - *S2:* a low wake-up latency state with a loss of CPU and system cache state.
    - *S3:* a low wake-up latency state in which all system state except the main memory is lost.
    - *S4:* the lowest power sleeping state, in which all devices are turned off.
- *G1:* The sleeping state in which the system appears to be off, and the time required to return to working condition is inversely proportional to power consumption.
- *G0:* The working state, in which the system is fully usable.
- *The legacy state:* The system does not comply with ACPI.

The power manager typically includes an observer that receives messages through the ACPI interface that describe the system behaviour. It also includes a decision module that determines the power management actions based on those observations.

11.4   Conclusion

In this chapter, we have seen different techniques for low power design of embedded systems. The power minimization can be targeted at different levels of design hierarchy. We have noted some of the prominent approaches at each level. System level power management issues, that lead to system monitoring and predictive shutdown to save power, have been discussed. The newer devices being manufactured are expected to support different power modes. The newer operating systems also include an ACPI module, so that the status of the devices can be monitored and appropriate power management commands can be issued to save system power.

Exercises

11.1 Why is power optimization an important issue in embedded system design? Enumerate various factors on which it depends.

11.2 Discuss about different sources of power dissipating in CMOS circuits.

11.3 Why is leakage power becoming increasingly important in submicron level design?

11.4 Mention the different levels at which power reduction can be attempted. From system power consumption view point, rank the levels on the basis of their potential to save power.

11.5 What are the algorithmic power minimization techniques?

11.6 How can parallelism and pipelining lead to architectural level power saving?

11.7 How does the dynamic power consumption differ in static and dynamic logic family? For the following truth-table, which corresponds to a majority gate function, obtain three different implementations—(i) AND-OR-INVERT, (ii) NAND, (iii) NOR. For each case, compute the total transition probabilities. Assume ON probabilities of all inputs to be 0.5.

a	b	c	o
0	0	0	0
0	0	1	0
0	1	0	0
0	1	1	1
1	0	0	0
1	0	1	1
1	1	0	1
1	1	1	1

11.8 How do the glitches contribute to power consumption? To reduce glitch, often redundant paths are added to a circuit. Is it also advisable from leakage viewpoint?

11.9 Decompose a 4-input OR gate in different ways using 2-input OR gates. Compare the total transition probabilities in the alternatives.

11.10 Explore different flip-flop structures targeted to low power dissipation.

11.11 Consider a 4-state finite-state machine with single input and single output. The state-transition table for the FSM is given below. Compare the power requirements (in terms of transition probabilities of combinational logic gates) for two different state assignments.

Present state	Primary input	Next state	Primary output
S0	0	S0	0
S0	1	S1	1
S1	0	S0	0
S1	1	S2	0
S2	0	S0	1
S2	1	S3	0
S3	0	S2	0
S3	1	S3	1

11.12 Justify the statement "clock-gating can reduce dynamic power only, whereas, power-gating can reduce both dynamic and leakage power".

11.13 What is predictive shutdown? How does it help in improving power performance of the system?

11.14 What is ACPI? What are the various power states of ACPI?

Bibliography

Alidina, M., Monteiro, J., Devadas, S., and Ghosh, A., Precomputation-Based Sequential Logic optimization for Low Power, *IEEE Transactions on Very Large Scale Integration (VLSI) Systems 2*, 4 (December 1994), 426–436.

Awad, M., Kuusela, J., and Ziegler, J., *Object Oriented Technology for Real-Time Systems*, Prentice Hall, 1996.

Balarin, F., Lavagno, L., Murthy, P., and Sangiovanni-Vincentelli, A., Scheduling for Embedded Real-Time Systems, *IEEE Design and Test of Computers* (1998), 71–82.

Ball, S., *Debugging Embedded Microprocessor Systems*, Newnes, 1998.

Ball, S., *Embedded Microprocessor Systems—Real World Designs*, Newnes, 1996.

Barnett, R., O'Cull, L., and Cox, S., *Embedded C programming and the Microchip PIC*, DELMAR Cengage Learning, 2004.

Barr, M., *Programming Embedded Systems*, O'Reilly, 1999.

Barrett, S.F., and Pack, D.J., *Embedded Systems Design and Applications with the 68HC 12 and HCS12*, Pearson Education, 2005.

Benini, L., and Michelli, G., *Dynamic Power Management—Design Techniques and CAD Tools*, Kluwer Academic Publisher, 1998.

Birkner, J., et al., A Very-High-Speed Field-Programmable Gate Array Using Metalto-Metal Antifuse Programmable Elements, *Microelectronics Journal 23* (1992), 561–568.

Bray, D.J., and Senese, B., *Bluetooth Applications Developer's Guide*, Syngress Publishing, 2001.

Burns, A., and Wellings, A., *Real-Time Systems and Programming Languages*, Addison-Wesley, 2001.

Buttazzo, G., *Hard Real-Time Computing Systems*, Kluwer Academic Publisher, 2002.

Camposano, R., and Brayton, R., Partitioning Before Logic Synthesis, *International Conference on Computer Aided Design*, 1987.

Catsoulis, J., *Designing Embedded Hardware*, O'Reilly, 2002.

Chandrakasan, A., Sheng, S., and Brodersen, R., Low Power CMOS Digital Design, *IEEE Journal of Solid-State Circuits 27*, 4 (1992), 119–123.

Chen, D., Sarrafzadeh, M., and Yeap, G., State Encoding of Finite State Machines for Low Power Design, *ISCAS'95* (1995), 2309–2312.

Chu, P.P., *FPGA Prototyping by VHDL Examples: XIlinx Spartan-3 Version*, Wiley-Interscience, 2008.

Chung, E., Benini, L., and Micheli, G., Source Code Transformation Based on Software Cost Analysis, *International Symposium on System Synthesis*, 2001, 153–158.

Chung, E.-Y., Benini, L., and Bogliolo, A., Dynamic Power Management for Non-stationary Service Requests, *IEEE Transactions on Computers 51*, 11 (November 2002), 1345–1361.

Debardelaben, J., Madisetti, V., and Gradient, A., Incorporating Cost Modeling into Embedded System Design, *IEEE Design and Test of Computers*, July 1997, 24–35.

Drusinsky, D., and Harel, D., Using StateCharts for Hardware Description and Synthesis, *IEEE Journal of Solid-State Circuits 27*, 4 (1989), 119–123.

Ernst, R., Henkel, J., and Benner, T., Hardware–Software Cosynthesis for Microcontrollers, *IEEE Design and Test of Computers* (December 1994), 64–75.

Fiduccia, C., and Mattheyses, R., A Linear-Time Heuristic for Improving Network Partitions, *Design Automation Conference*, 1982.

Furber, S., *ARM System-on-chip Architecture*, Addison-Wesley.

Gajski, D., and Vahid, F., Specification and Design of Embedded Hardware–Software Systems, *IEEE Design and Test of Computers*, 12, 1 (1995), 53–67.

Gajski, D., Vahid, F., Narayan, S., and Gong, J., *Specification and Design of Embedded Systems*, Prentice Hall, 1994.

Gajski, D., Vahid, F., Narayan, S., and Gong, J., Specsyn: An Environment Supporting the Specify-Explore-Refine Paradigm for Hardware/Software System Design, *IEEE Transactions on VLSI*, 1998.

Ganssle, J., *The Art of Designing Embedded Systems*, Newnes, 2000.

Gilster, D.M., *Bluetooth End to End*, John Wiley & Sons, 1999.

Gofton, P., *Mastering Serial Communications*, SYBEX, 1994.

Gomaa, H., *Software Design Methods for Concurrent and Real-time Systems*, Addison-Wesley, 1993.

Greef, E.D., Catthoor, F., and Man, H., Array Placement for Storage Size Reduction in Embedded Multimedia Systems, *IEEE International Conference on Application-Specific Systems*, Architectures and Processors (1997), 66–75.

Grout, I., *Digital Systems Design with FPGAS*, Newnes, 2008.

Gupta, R., Special Issue: Partitioning Methods for Embedded Systems, *Journal of Design Automation of Embedded Systems 2*, 2 (1997).

Gupta, R., and DeMicheli, G., Partitioning of Functional Models of Synchronous Digital Systems, *International Conference on Computer Aided Design,* 1990, 216–219.

Gupta, R., and DeMicheli, G., Hardware-Software Cosynthesis for Digital Systems, *IEEE Design and Test of Computers* (October 1993), 29–41.

Harel, D., StateCharts: A Visual Formalism for Complex Systems, *Science of Computer Programming* (1987), 231–274.

Heath, S., *Embedded Systems Design,* Elsevier, 2003.

Henkel, J., and Ernst, R., A Hardware/Software Partitioner using a Dynamically Determined Granularity, *Design Automation Conference,* 1997.

Hohl, W., *ARM Assembly Language: Fundamentals and Techniques,* CRC Press, 2009.

Hou, J., and Wolf, W., Process Partitioning for Distributed Systems, *CODES* (1996), 70–75.

Hyde, J., *USB Design by Example,* Wiley, 2002.

Johannes, F., Partitioning of VLSI Circuits and Systems, *Design Automation Conference,* 1996.

Kalavade, A., and Lee, E., A Global Criticality/Local Phase Driven Algorithm for the Constrained Hardware/Software Partitioning Problem, *CODES* (1994), 42–48.

Kamal, R., *Embedded Systems Architecture, Programming and Design,* Tata McGraw Hill, 2008.

Keding, H., Willems, M., Coors, M., and Meyr, H., FRIDGE: A Fixed-Point Design and Simulation Environment, *Design Automation and Test in Europe* (DATE) (1998), 429–435.

Kernighan, B., and Lin, S., An Efficient Heuristic Procedure for Partitioning Graphs, *Bell System Technical Journal* (February 1970).

Knudsen, P., and Madsen, J., PACE: A Dynamic Programming Algorithm for Hardware/ Software Partitioning, *International Workshop on Hardware–Software Codesign,* (1996), 85–92.

Krishna, C., and Shin, K., *Real-Time Systems,* Tata McGraw-Hill, 1997.

Kuon, I., Tessier, R., and Rose, J., *FPGA Architecture Survey and Challenges,* Now Publisher.

Lagnese, E., and Thomas, D., Architectural Partitioning for System Level Synthesis of Integrated Circuits, *IEEE Transactions on Computer-Aided Design 10* (July 1991), 847–860.

Laplante, P., *Real-Time Systems Design and Analysis: An Engineer's Handbook,* IEEE Press, 1997.

Lapsley, P., Bier, J., Shoham, A., and Lee, E.A., *DSP Processor Fundamentals: Architectures and Features,* Wiley-IEEE Press, 1997.

Lehoczky, J.L., and Sha, L., The Rate-Monotonic Scheduling Algorithm: Exact Characterization and Average Case Behaviour, *Proceedings of Real-Time Systems Symposium* (December 1989), 166–171.

Lewis, D.W., *Fundamentals of Embedded Software*, PHI Learning, 2002.

Liu, C., *Real-Time Systems*, Prentice Hall, 2000.

Liu, C., and Layland, J. Scheduling Algorithms for Multi-Programming in a Hard Real-Time Environment, *Journal of Association of Computing Machinery* (1973), 40–71.

Mall, R., *Real-Time Systems Theory and Practice*, Pearson-Education, 2007.

Marwedel, P., *Embedded System Design*, Springer, 2006.

Mazidi, M.A., Mazidi, J.G., and McKinlay, R.D., *The 8051 Microcontroller and Embedded Systems*, PHI Learning, 2006.

Morton, T.D., *Embedded Microcontrollers*, Pearson Education, 2001.

Moyer, B., Low-Power Design for Embedded Processors, Proceedings of the IEEE 89, 11 (November 2001), 1576–1587.

Raghunathan, A., Dey, S., and Jha, N., Register Transfer Level Power Optimization with Emphasis on Glitch Analysis and Reduction, *tcad* (1999), 1114–1131.

Ramamritham, K., and Stancovic, J., Scheduling Algorithms and Operating Systems Support for Real-Time Systems, *Proceedings of the IEEE* (January 1994), 55–67.

Seal, D., *ARM Architecture Reference Manual*, Addision-Wesley Professional, 2001.

Sha, L., Rajkumar, R., and Lehoczky, J., Priority Inheritance Protocols: An Approach to Real-Time Synchronization, *IEEE Transactions on Computers*, 1990, 1175–1185.

Shi, C., and Brodersen, R., An Automated Floating-Point to Fixed-Point Conversion Methodology, *International Conference on Audio Speech and Signal Processing*, 1995, 529–532.

Sloss, A., Symes, D., and Wright, C., *ARM Architecture Reference Manual*, Morgan Kauffman, 2004.

Sloss, A., Symes, D., and Wright, C., *ARM System Developer's Guide: Designing and Optimizing System Software*, Morgan Faufmann, 2004.

Smith, S.W., *The Scientist and Engineer's Guide to Digital Signal Processing*, California Technical Publishing.

Stankovic, J., Spuri, M., Ramamritham, K., and Buttazzo, G., *Deadline Scheduling for Real-Time Systems, EDF and Related Algorithms*, Kluwer Academic Publishers, 1998.

Staunstrup, J., and Wolf, W., *Hardware/Software Co-Design Principles and Practice*, Springer, 1997.

Strollo, A., Napoli, E., and Caro, D.D., New Clock-Gating Techniques for Low Power Flip-Flops, *International Symposium on Low Power Electronics and Design* (July 2000), 114–119.

Tiwari, V., Ashar, P., and Malik, S., Technology Mapping for Low Power, *30th ACM/IEEE Design Automation Conference*, 1993, 68–73.

Tsui, C., Pedram, M., and Despain, A., Technology Decomposition and mapping targeting low power dissipation, *30th ACM/IEEE Design Automation Conference*, (1993), 68–73.

Vahid, F., Procedure Exlining: A Transformation for Improved System and Behavioural Synthesis, *International Symposium on System Synthesis* (1995), 84–89.

Vahid, F., Port Calling: A Transformation for Reducing I/O During Multipackage Functional Partitioning, *International Symposium on System Synthesis*, 1997.

Vahid, F., Procedure Cloning: A Transformation for Improved System-Level Functional Partitioning, *European Design and Test Conference*, 1997, 487–492.

Vahid, F., and Gajski, D., Closeness Metrics for System-Level Functional Partitioning, *European Design Automation Conference*, 1995, 328–333.

Vahid, F., and Gajski, D., Clustering for Improved System-Level Functional Partitioning, *International Symposium on System Synthesis*, 1995, 28–33.

Vahid, F., and Gajski, D., Incremental Hardware Estimation During Hardware/Software Functinal Partitioning, *IEEE Transactions on Very Large Scale Integration Systems 3*, 3 (1995), 459–464.

Vahid, F., and Givargis, T., *Embedded System Design A Unified Hardware/Software Introduction*, Wiley, 2002.

Vahid, F., and Le, T., Towards, A Model for Hardware and Software Functional Partitioning, *International Workshop on Hardware–Software Codesign* (1996), 116–123.

Vahid, F., and Le, T., Extending the Kernighan-Lin Heuristic for Hardware and Software Functional Partitioning, *Journal of Design Automation of Embedded Systems 2*, 2 (1997), 237–261.

Vahid, F., Le, T., and Hsu, Y., A Comparison of Functional and Structural Partitioning, *International Symposium on System Synthesis*, 1996, pp. 121–126.

Valvano, J.W., *Embedded Microcomputer Systems*, Cengage Learning, 2007.

Wolf, W., Hardware-Software Codesign of Embedded Sytems, Proceedings of the IEEE 82, 7 (1994), 967–989.

Wolf, W., *Computers as Components–Principles of Embedded Computing System Design*, Elsevier, 2008.

Xue, J., *Loop Tiling for Parallelism*, Kluwer Academic Publishers, 2000.

Index